T0210760

EVOLUTION OF PHASE TRANSITIONS

This work began with the authors' exploration of the applicability of the finite deformation theory of elasticity when various standard assumptions such as convexity of the energy or ellipticity of the field equations of equilibrium are relinquished. The finite deformation theory of elasticity turns out to be a natural vehicle for the study of phase transitions in solids where thermal effects can be neglected. This is a valuable work for those interested in the development and application of continuum-mechanical models that describe the macroscopic response of materials capable of undergoing stress- or temperature-induced transitions between two solid phases. The focus is on the evolution of phase transitions, which may be either dynamic or quasi-static, controlled by a kinetic relation that in the framework of classical thermomechanics represents information that is supplementary to the usual balance principles and constitutive laws of conventional theory. The book should be of interest to mechanicians, material scientists, geophysicists, and applied mathematicians.

Rohan Abeyaratne is the Quentin Berg Professor of Mechanics and Head of the Department of Mechanical Engineering at MIT. He received his bachelor's degree from the University of Ceylon and his doctorate from the California Institute of Technology. Among his honors are the E.O.E. Pereira Gold Medal (1975), Den Hartog Distinguished Educator (1995), MacVicar Fellowship (2000), Fellow, American Academy of Mechanics (1996) and Fellow, American Society of Mechanical Engineers (1998). His primary research interest is in nonlinear phenomena in mechanics.

James K. Knowles is the William R. Kenan Professor of Applied Mechanics, Emeritus, at the California Institute of Technology. He received his S.B. and Ph.D. degrees from MIT, and he holds an honorary Sc.D. degree from the National University of Ireland. He is a Fellow of the American Academy of Mechanics (AAM), the American Association for the Advancement of Science and the American Society of Mechanical Engineers (ASME). He is a past president of AAM, and he is a recipient of MIT's Goodwin Medal for teaching, the Eringen Medal of the Society of Engineering Science and the Koiter Medal of the ASME. His primary research interests are in nonlinear phenomena in continuum mechanics, and in analytical issues in fracture mechanics and the theory of elasticity.

EVOLUTION OF PHASE TRANSITIONS

TRANSITIONS

A Continuum Theory

ROHAN ABEYARATNE

Massachusetts Institute of Technology

JAMES K. KNOWLES

California Institute of Technology

CAMBRIDGE
UNIVERSITY PRESS

CAMBRIDGE UNIVERSITY PRESS
Cambridge, New York, Melbourne, Madrid, Cape Town,
Singapore, São Paulo, Delhi, Tokyo, Mexico City

Cambridge University Press
32 Avenue of the Americas, New York, NY 10013-2473, USA

www.cambridge.org
Information on this title: www.cambridge.org/9780521380515

© Cambridge University Press 2006

This publication is in copyright. Subject to statutory exception
and to the provisions of relevant collective licensing agreements,
no reproduction of any part may take place without the written
permission of Cambridge University Press.

First published 2006
First paperback edition 2011

A catalog record for this publication is available from the British Library

Library of Congress Cataloging in Publication data

Abeyaratne, Rohan.
Evolution of phase transitions : a continuum theory / Rohan Abeyaratne, James K. Knowles.
 p. cm.
Includes bibliographical references and index.
ISBN-13: 978-0-521-66147-8
ISBN-10: 0-521-66147-1
1. Phase transformations (Statistical physics). 2. Continuum mechanics.
3. Kinetic theory of matter. 1. Knowles, James K. (James Kenyon), 1931–
11. Title
QC175.16.P5A24 2006
530.4′74—dc22 2005033285

ISBN 978-0-521-66147-8 Hardback
ISBN 978-0-521-38051-5 Paperback

Cambridge University Press has no responsibility for the persistence or
accuracy of URLs for external or third-party internet websites referred to in
this publication, and does not guarantee that any content on such websites is,
or will remain, accurate or appropriate.

To the C7: Gina, Kenny, Kevin, Kristen, Liam, Linus, & Nina;

and the J4: Jackie, John, Jeff, & Jamey.

Contents

**Part IV One-Dimensional Thermoelastic Theory
          and Problems**

Preface

This monograph threads together a series of research studies carried out by the authors over a period of some fifteen years or so. It is concerned with the development and application of continuum-mechanical models that describe the macroscopic response of materials capable of undergoing stress- or temperature-induced transitions between two solid phases.

Roughly speaking, there are two types of physical settings that provide the motivation for this kind of modeling. One is that associated with slow mechanical or thermal loading of alloys such as nickel–titanium or copper–aluminum–nickel that exhibit the shape-memory effect. The second arises from high-speed impact experiments in which metallic or ceramic targets are struck by moving projectiles; the objective of such studies – often of interest in geophysics – is usually to determine the response of the impacted material to very high pressures. Phase transitions are an essential feature of the shape-memory effect, and they frequently occur in high-speed impact experiments on solids. Those aspects of the theory presented here that are purely phenomenological may well have broader relevance, in the sense that they may be applicable to materials that transform between two "states," for example, the ordered and disordered states of a polymer.

Our development focuses on the evolution of the phase transitions modeled here, which may be either dynamic or quasistatic. Such evolution is controlled by a "kinetic relation," which, in the framework of classical thermomechanics, represents information supplementary to the usual balance principles and constitutive laws of conventional theory. We elucidate the rather remarkable way in which the classical theory "calls for" this kind of supplementary information when the material is capable of changing phase, though such additional information is *not* called for – indeed, cannot be imposed – in the case of a single-phase material.

The simplest context in which to illustrate the need for kinetic relations and the role they play is that furnished by the purely mechanical theory of one-dimensional nonlinear elasticity, with thermal effects suppressed. After the Introduction, which comprises Part I of the monograph, we pursue the subject in this context in Part II. Even this simplest version of the theory to be set out here has some utility, as we show in Chapters 3 and 4. Part III presents the full three-dimensional theory, taking both mechanical and thermal effects into account. We specialize this theory to one space dimension in Part IV, where we are able to make some comparisons with experiments. In Part V, we discuss some three-dimensional problems.

The material presented here is drawn primarily from our own research over the period from the late 1980s forward. We came to this subject as practitioners of solid mechanics interested in exploring the range of applicability of the finite deformation theory of elasticity when various standard assumptions such as convexity of various energies or ellipticity of the field equations of equilibrium were relinquished. When broadened in this way, finite elasticity is a natural vehicle for the study of those aspects of phase transitions in solids that can be discussed with thermal effects neglected. Nonlinear *thermoelasticity*, similarly unencumbered by conventional restrictions, provides the natural framework for the study of mechanical and thermal effects together.

Our hope is that this book will be of interest to materials scientists, engineers and geophysicists as well as to mechanicians and applied mathematicians. The perfectly prepared reader would be acquainted with continuum mechanics at the level of Chadwick's *Continuum Mechanics*, Wiley, New York, 1976; with thermodynamics as treated, for example, in J. L. Ericksen's *Introduction to the Thermodynamics of Solids*, Chapman and Hall, New York, 1991; with material behavior as described by T. H. Courtney in *Mechanical Behavior of Materials*, McGraw-Hill, New York, 1990; with partial differential equations at the level of J. D. Logan's *An Introduction to Nonlinear Partial Differential Equations*, Wiley-Interscience, New York, 1994; and with the elements of Cartesian tensors as discussed, for example, in *Linear Vector Spaces and Cartesian Tensors*, Oxford, New York, 1998, by J. K. Knowles. However, expecting many potential readers to be less than perfectly prepared, we have tried to make the presentation as self-contained as is practicable, citing appropriate sources for those results that are used but not derived.

Although the book deals almost entirely with our own work, we have nevertheless had the enormous benefit of interactions with many others, and it is a pleasure to acknowledge them *all* with gratitude. We would be remiss not to mention the particular influence that Tom Ahrens, Kaushik Bhattacharya, Mort Gurtin, Rick James, Stelios Kyriakides, Jim Rice, the late Eli Sternberg, Lev Truskinovsky, and our former doctoral students, especially Phoebus Rosakis and Stewart Silling, have had on our learning of this subject.

Some of the fruitful interactions alluded to above took place in small, informal summer gatherings held at MIT's Talbot House in South Pomfret, Vermont. We are indebted to MIT for the use of this wonderful place, which – alas – is no longer owned by MIT.

Special thanks go to Debbie Blanchard, who drew the figures in the early part of the book, and then taught us how to draw the rest.

We are grateful to Olaf Weckner for a careful and constructive critical reading of the early chapters.

We acknowledge with thanks the past financial support of the U.S. National Science Foundation, the U.S. Army Research Office, and especially the U.S. Office of Naval Research, with which we enjoyed a sustained relationship and which supported much of the research on which this monograph is based. We would particularly like to thank Roshdy Barsoum, Alan Kushner, and Yapa Rajapakse for the help and encouragement that they, as program officers at ONR, consistently provided to us.

During recent stimulating visits, both of us have benefited from the hospitality and financial support of the University of Cambridge, its colleges, and its Isaac Newton Institute for the Mathematical Sciences, for which we wish to express our appreciation.

Rohan Abeyaratne and Jim Knowles
Cambridge, Massachusetts, and Pasadena, California
June 2005

Part I Introduction

1 Introduction

1.1 What this monograph is about

Certain crystalline materials can exist in more than one solid phase, where a phase is identified by a distinct crystal structure. Typically, one phase is preferred under certain conditions of stress and temperature, while another is favored under different conditions. As the stress or temperature varies, the material may therefore transform abruptly, from one phase to another, leading to a discontinuous change in the properties of the body. Examples of such materials include the shape-memory alloy NiTi, the ferroelectric alloy $BaTiO_3$, the ferromagnetic alloy FeNi and the high-temperature superconducting ceramic alloy $ErRh_4B_4$. In each of these examples the transition occurs without diffusion and one speaks of the transformation as being *martensitic* (or displacive).

Alloys such as Au–47.5%Cd and Cu–15.3%Sn are known to have a cubic lattice at high temperatures and an orthorhombic lattice at low temperatures. Therefore, if such a material is subjected to thermal cycling, it will transform between these two phases. Similarly, alloys such as Ni–36%Al and Fe–7%Al–2%C transform between a high-temperature cubic phase and a low-temperature tetragonal phase, whereas near-equiatomic NiTi has a high-temperature cubic phase and low-temperature monoclinic phase.

If a stress-free single crystal of such a two-phase material is slowly cooled from a sufficiently high temperature, it starts out in the high-temperature phase and at first, merely undergoes a thermal contraction. However at some critical temperature, denoted M_s in the materials science literature[1], a portion of the lattice suddenly transforms from the high-temperature structure to the low-temperature structure. Most of the specimen is still in the high-temperature phase, but a small amount is now in the low-temperature phase. The interface between the regions occupied by the two phases is a "phase boundary." As the temperature is further decreased, the phase boundary propagates into the high-temperature phase, thus transforming

---

[1] The subscripts s and f denote "start" and "finish" respectively, whereas M and A denote martensite, the low-temperature phase, and austenite, the high-temperature phase. So, for example, M_s is the temperature at which the formation of <u>m</u>artensite <u>s</u>tarts, etc.

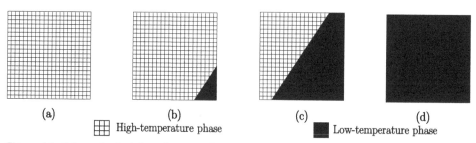

Figure 1.1. Schematic depiction of a stress-free single crystal of a two-phase material at a sequence of progressively decreasing temperatures.

it into the low-temperature phase as it crosses the moving front. Eventually, at some temperature M_f, the entire specimen is in the low-temperature phase. Further cooling simply causes thermal contraction of the low-temperature phase. This is depicted in Figure 1.1 by the sequence of diagrams (a)→(b)→(c)→(d).

If the specimen is now gradually reheated, the reverse process takes place. The high-temperature phase is nucleated at some temperature A_s, it grows by the propagation of phase boundaries, and finally the entire specimen returns to the high-temperature phase at a temperature A_f. Further heating simply causes thermal expansion.

If during the aforementioned cooling–heating cycle one plots some representative linear dimension ℓ of the specimen versus the temperature θ, one obtains a hysteresis loop as shown schematically in Figure 1.2. The cooling portion of the cycle corresponds to ① → ② → ③ → ④ → ⑤. Figure 1.1(a) corresponds to the initial cooling segment ① → ② → ③ during which the entire specimen is in the high-temperature phase. The low-temperature phase is nucleated at ③ and grows from ③ → ④ as the specimen evolves from (a)→(b)→(c)→(d) in Figure 1.1. The transformation is complete at ④ (Figure 1.1(d)) and thereafter undergoes pure thermal contraction from ④ → ⑤. The reverse transformation ⑤ → ④ → ⑥ → ② → ① occurs during reheating.

If, instead of being stress free, the specimen was held at some constant nonzero level of stress (in a suitable range) during the thermal cycling, one observes a similar sequence of events but the quantitative details will be different. Likewise, if instead of thermal loading, the specimen is subjected to a cyclic mechanical loading, one again observes a qualitatively similar sequence of events. One can consider a wide range of loading rates, either thermal or mechanical, with the associated response ranging from quasistatic to inertia driven, and isothermal to adiabatic, and in each case (provided one stays in a suitable range) one can induce forward and reverse phase transitions, the detailed response depending on the specifics of the loading conditions.

A conceptually helpful way in which to think about the modeling of a martensitic phase transition using the continuum theory of finite thermoelasticity is as follows: a general thermoelastic material is characterized by its free energy potential $\psi(\mathbf{F}, \theta)$, which is a function of the deformation gradient tensor $\mathbf{F}$ and the

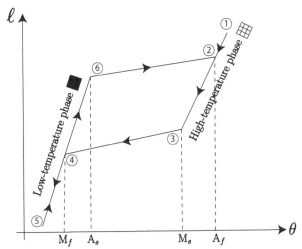

Figure 1.2. A schematic plot of a representative linear dimension ℓ of the specimen versus temperature θ during a cycle of cooling and reheating. The specimen transforms between a high-temperature phase and a low-temperature phase.

absolute temperature θ. The particular form of $\psi(\mathbf{F}, \theta)$ corresponding to a given material is determined by the microstructural details of that material. Consider a given thermoelastic body at a temperature θ. Suppose that at this θ, the potential ψ as a function of $\mathbf{F}$ has a local minimum at $\mathbf{F} = \mathbf{F}^+$. Then, since the gradient of ψ with respect to $\mathbf{F}$ is essentially the stress, the pair $(\mathbf{F}^+, \theta)$ describes a homogeneous stress-free equilibrium state of this body in the sense that if the entire body is homogeneously deformed by the deformation gradient tensor $\mathbf{F}^+$, and the temperature is held at θ, then the body will be stress free and can therefore be maintained in equilibrium without the application of any external loading.

If $\psi(\cdot, \theta)$ happens to have *two* local minima, say at $\mathbf{F}^+$ and $\mathbf{F}^-$ as depicted schematically in Figure 1.3(a), and if $\mathbf{F}^+$ and $\mathbf{F}^-$ do not differ by a rigid rotation, then there are *two* distinct stress-free homogeneous equilibrium states $(\mathbf{F}^+, \theta)$ and $(\mathbf{F}^-, \theta)$, *either of which* the body can occupy. Typically, each of these equilibrium states is associated with a distinct *phase* of the material.

Under certain circumstances the two phases can coexist, with part of the body being in the homogeneous stress-free equilibrium state $(\mathbf{F}^+, \theta)$, and the remainder being in the homogeneous stress-free equilibrium state $(\mathbf{F}^-, \theta)$; the interface separating these two parts of the body is a phase boundary. Since each part of the body is stress free, this piecewise homogeneous configuration will be in mechanical equilibrium without the application of any external loading.

If the values $\psi^+ = \psi(\mathbf{F}^+, \theta)$ and $\psi^- = \psi(\mathbf{F}^-, \theta)$ of the free energy associated with the two phases are equal, then a configuration that involves both phases is in *phase equilibrium* as well as in mechanical equilibrium: each phase has the same energy and therefore neither is preferred over the other. On the other hand if, for example, $\psi^+ > \psi^-$, as in Figure 1.3(a), then the phase associated with $\mathbf{F}^-$ is the low-energy phase and is therefore energetically favored over the other phase.

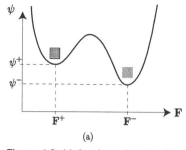
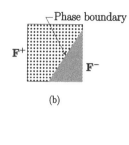

Figure 1.3. (a) A schematic plot of the free energy ψ versus the deformation gradient **F** at a fixed temperature, showing local minima at $\mathbf{F} = \mathbf{F}^+$ and $\mathbf{F} = \mathbf{F}^-$. (b) Two phases characterized by $\mathbf{F}^+$ and $\mathbf{F}^-$ coexisting in a body, separated by a phase boundary.

There is now a "thermodynamic driving force," which tends to transform the high-energy phase "plus" into the low-energy phase "minus". Accordingly, this driving force makes it favorable for the phase boundary to move into the region associated with $\mathbf{F}^+$, thereby transforming the high-energy phase into the low-energy phase as it crosses the moving interface. The driving force vanishes in phase equilibrium. How fast the phase boundary moves – the question of "kinetics" – is determined by the microstructure underlying the two phases and the details of the mechanism by which one lattice changes into the other.

In order to relate this discussion to the earlier one, we simply need to relate the plus and minus phases here to what we previously called the high-temperature and low-temperature phases. For example, if, as a function of the temperature θ, the energies $\psi(\mathbf{F}^+, \theta)$ and $\psi(\mathbf{F}^-, \theta)$ are such that $\psi^+ > \psi^-$ for high temperatures and $\psi^+ < \psi^-$ for low temperatures, then the minus phase is energetically preferred at high temperatures and can therefore be identified with what we previously called the high-temperature phase; similarly the plus phase can be identified with the low-temperature phase.

Next, consider a stress-free body composed of two coexistent phases that are in mechanical and phase equilibrium. If the body is now subjected to some stress, this alters the relative favorability of the two phases and will, in principle, lead to phase boundary motion. Similarly, this will happen if the temperature is changed.

It should be noted that in the *phenomenological* theory of thermoelasticity, if one is simply given a potential ψ with multiple local minima, one cannot say much about the different phases associated with the different minima. They might, for example, be associated with the liquid and gaseous phases of an elastic fluid, two crystallographically distinct lattices of a crystalline solid, the disordered and ordered states of a polymer, and so on. In order to physically identify the phases one must consider the underlying microstructure.

This monograph explores in detail a number of the concepts referred to above. We study questions related to the quasistatic and dynamic responses of an elastic material characterized by a free energy potential with multiple local minima. We pay particular attention to the notions of *driving force* and *kinetics*. The driving force plays a central role in characterizing the dissipation rate during time-dependent processes, and we make this notion rigorous under very general circumstances and then specialize it to various particular cases. We address both modeling and mathematical issues. For example on the modeling front, we calculate an explicit free energy function ψ for a material with cubic and tetragonal phases. We develop various models, both phenomenological and micromechanical, of the transformation kinetics. And we solve a number of boundary-initial value problems connected to experiments. On the mathematical front, we show that quasistatic and dynamic problems for two-phase materials, when formulated as in classical elasticity, suffer from a severe lack of uniqueness of solution to boundary–initial value problems. We show that the degree of nonuniqueness is *precisely* filled up by a proper description of the nucleation and kinetics of the transformation.

This monograph is not meant to provide an encyclopedic treatment of phase transitions. Rather, it simply collects together, and retells as a single story, the results from a series of studies mostly by the present authors on the motion of phase boundaries in solids.

1.2 Some experiments

The article by Schetky [45] provides a general introduction to the subject of martensitic transformations while the book by Duerig et al. [18] describes many engineering applications. Comprehensive overviews of the materials science literature on this subject can be found in Christian [14], Funakubo [25], and Otsuka and Wayman [42] among others

Experiments that investigate the thermomechanical response of martensitic materials can, roughly, be grouped into three sets: uniaxial tensile loading experiments, multiaxial loading experiments, and high-rate experiments.

There is an extensive experimental literature devoted to tensile loading and unloading of bars made of martensitic materials; often, the materials studied are technologically important shape-memory alloys such as NiTi. For a small sample of this literature, see, for example, Gao and Brinson [26], Krishnan and Brown [32], Nakanishi [39], Shaw and Kyriakides [46], and Leo *et al.* [36]. The loading in such experiments is slow, in the sense that inertia is insignificant. Some of these experiments are so slow that the process is essentially isothermal; but others, such as those of Shaw and Kyriakides [46] and Leo *et al.* [36], are sufficiently fast that the effects of local heating/cooling at the phase boundary become significant. The objective of these experiments is typically to determine the relation between the applied stress and the overall elongation of the bar, though in some studies such as that of Shaw and Kyriakides [46], local strain and temperature measurements along the length of the specimen are also made. The observed stress–elongation relation exhibits hysteresis, and the phase transition represents the primary mechanism responsible for this dissipative behavior. For a given material, the size and other features of the hysteresis loop depend on the loading rate and the temperature at which the test takes place.

Investigations on the multiaxial loading of martensitic materials are not as numerous. Among the comparatively few that do exist are the tension-torsion experiments of Lim and McDowell [37] and the biaxial tensile loading experiments of Chu and James [1, 15]. During a typical test, two stress components are prescribed as functions of time and the response of the conjugate kinematic quantities is measured, or vice versa, resulting, again, in a hysteretic response.

There is a large body of experimental literature describing the response of solids to shock- or impact-loading. Much of it is motivated by questions concerning the behavior of materials at extremely high pressures, as occurs, for example, deep in the earth; see for example, Meyers [38]. In a typical experiment, a specimen in the form of a cylindrical disk is subjected to a high-speed normal impact on one of its plane faces. The collision generates a compressive disturbance that propagates into the disk, ultimately reaching the rear face of the target. Propagation speeds of the disturbances involved are measured, as are particle velocities at the rear face. Depending on the material, a phase transition may occur, as, for example, in the case of the graphite-to-diamond transition studied by Erskine and Nellis [20], or the austenite-to-martensite transition in CuAlNi studied by Escobar and Clifton [21]. One goal of such experiments is to determine the relation between the impact speed and the resulting speed of the phase transformation.

Some of the aforementioned experiments have been carried out on polycrystalline specimens of the material, while others use single crystals. Experiments on single crystals usually aim to understand the local processes associated with the propagation of phase boundaries and to relate this to the macroscopic response, for example, Chu and James [1, 15], Gao and Brinson [26].

In most martensitic materials, one does *not* observe a simple piecewise homogeneous deformation during phase transformation involving two homogeneously deformed phases separated by a planar phase boundary. Usually, the crystallographic structure on one side of a phase boundary is more complicated than this. In some cases, for example, In–22.73%Tl, one observes a homogeneously deformed high-temperature phase separated by a planar phase boundary from a low-temperature phase that has a "twinned" fine structure. Another common microstructure involves a wedge of the twinned low-temperature material surrounded by the homogeneously deformed high-temperature phase, seen, for example, in Ni–36%Al. However there do exist some exceptional materials, such as some of the titanium–tantalum alloys studied by Bywater and Christian [13], which do exhibit simple piecewise homogeneous deformations during phase transformation. As will be illustrated in Chapter 12, the ability to form an interface between the high-temperature and low-temperature phases without twinning requires that the lattice parameters of the two phases be related in special ways. It is conceivable that one could adjust the composition of a shape memory alloy so as to make the lattices satisfy these special relationships, and therefore to make the material exhibit phase boundaries that do not involve finely twinned structures.

1.3 Continuum mechanics

In 1975, Ericksen [19] examined the equilibrium of an elastic bar composed of a material with a nonmonotonic stress–strain relation. The strain-energy function associated with this stress–strain relation had two local minima and could therefore be used to describe a two-phase material. Following on this seminal paper, there has been a vast amount of activity in the continuum mechanics community on the modeling of solid–solid phase transitions.

Work on this subject can broadly be divided into two parts. One set of studies has been focused on examining equilibrium states of a two-phase solid using energy minimization as the underlying principle. The richness of this subject stems from the fact that, as mentioned previously, observed microstructures can be quite complicated. The low-temperature phase often involves a fine structure ("twinning") and in some cases even involves a fine structure within a fine structure. Understanding and modeling these complex microstructures has been a challenging and rewarding subject. The recent book by Bhattacharya [10] and the earlier review article by James and Hane [29] provide comprehensive discussions of the theory of equilibrium microstructures.

The second set of studies concerns the time-dependent response of these materials corresponding to various mechanical and/or thermal loading processes. At the microscale, the goal is to understand the evolution of the microstructures, while at the macroscale one wishes to describe various aspects of the hysteretic macroscopic response. Internal-variable-based models have been developed and explored by many investigators, for example, Brinson, Lagoudas, Müller, and co-workers (e.g. [12, 11, 8]). Phase field models have also been successfully used by others including Fried, Khachaturyan, Voorhees, and co-workers, for example, [24, 9, 49]. Models that use viscosity to capture dissipation and a strain-gradient term to enable propagation have been particularly popular and feature prominently in the work of, for example, Shearer [47], Slemrod [48], and Truskinovsky [51].

Irrespective of the details of a specific material system, a hysteretic response involves energy dissipation, and one of the fundamental questions pertains to how one characterizes dissipation in a system where the primary source of dissipation is phase transformation. Knowles and Sternberg [31] and Knowles [30] showed that, much like a propagating shock wave in an inviscid compressible fluid, or a propagating crack in an elastic solid, the strain discontinuity that occurs across a propagating phase boundary can be a source of dissipation even in slow processes. This, coupled with the basic notions of irreversible thermodynamics, leads one to the notion of the "driving force" on a phase boundary – the primary agent responsible for entropy production due to phase transformation; see Abeyaratne and Knowles [2, 6], Heidug and Lehner [28], Truskinovsky [50]. It is related to the general concept of configurational forces as developed by Gurtin [27], as well as to notions introduced much earlier by Eshelby [22]. Thus, during the process of phase transformation, there is a driving force on a phase boundary, and as material crosses this moving interface, it transforms from the high-energy phase to the low-energy phase. The flux that is conjugate to this driving force is the rate at which the material transforms, or equivalently, the (Lagrangian) speed of the interface. The process of phase transformation, like any nonequilibrium process, is governed by a kinetic law that relates the flux to the driving force and other appropriate variables. Therefore in order to fully model the time evolution of a phase transition, one must specify an elastic potential that characterizes each phase and a kinetic law that characterizes the evolution of the phases. The detailed form of the kinetic law depends on the microscale dynamic process by which one lattice transforms to the other. Some preliminary efforts to derive kinetic models from lattice considerations can be found in Abeyaratne and Vedantam [7], and Purohit [44].

1.4 Quasilinear systems

The second-order quasilinear system of partial differential equations

$$\frac{\partial h(u_1)}{\partial x} = \frac{\partial u_2}{\partial t}, \quad \frac{\partial u_2}{\partial x} = \frac{\partial u_1}{\partial t}$$

for two unknowns $u_1(x, t)$, $u_2(x, t)$ has been extensively studied in the applied mathematics literature, for example [23]. This system arises in many physical contexts, for example, in gas dynamics, it describes the flow of a compressible inviscid fluid in which case one identifies u_1 with mass density, u_2 with particle velocity, and $h(u_1)$ with the function characterizing the pressure–density relation of the gas. Likewise, by identifying u_1 with strain, u_2 with particle velocity and $h(u_1)$ with the function that describes the stress–strain relation of a solid, it can be used to describe the dynamics of a nonlinearly elastic bar.

In gas dynamics, it is well known that the Cauchy problem for this system suffers from nonuniqueness of solution, but that there is a unique weak solution that also satisfies the associated entropy inequality. More generally, for any function h that is either a strictly convex or a strictly concave function of u_1, the Cauchy problem for this system can be shown to have at most one weak solution fulfilling a so-called "admissibility criterion", Oleinik [40]. On the other hand, when h is not strictly convex or strictly concave, the Cauchy problem need not have a unique solution, *even with* the entropy inequality in force, Dafermos [17].

The issue of selecting a unique solution to a quasilinear system in the case where h is not a strictly convex or strictly concave function of u_1 has been addressed by many authors, usually by replacing the entropy inequality by a stronger "admissibility criterion," for example Dafermos [16], Lax [33], and LeFloch [35]. Among the most common such admissibility criteria are the viscosity criterion, the viscosity–capillarity criterion, the entropy-rate admissibility criterion, and the chord criterion, for example [43, 48, 16, 41].

The elastic bar considered by Ericksen [19] in his study of a two-phase material can be described by a function h that, with increasing strain u_1, first increases, then decreases, and finally increases again. Such a function h is neither strictly convex nor strictly concave, and so the quasilinear system suffers from the lack of uniqueness alluded to above, even with the entropy inequality in place. Rather than replacing the entropy inequality with a "uniqueness-generating admissibility criterion," we take the point of view that the lack of uniqueness reflects a deficiency in the physical model. The entropy inequality, being a statement of the second law of thermodynamics, must be retained. Rather, additional information from the physics of phase transformations needs to enter the mathematical model. Important insight into the information that is lacking can be obtained if one can determine *all* solutions to the quasilinear system and organize them in some illuminating way. We were able to do this explicitly in a special, algebraically simple case, [3], and thereby conclude that the nonuniqueness arises for two distinct reasons: it pertains to (i) the question of whether or not the motion involves a phase boundary and (ii) if the motion does involve a phase boundary, then the indeterminacy of the phase boundary speed. This suggests, exactly as the materials scientists would tell us, that the quasilinear system must be supplemented with a "nucleation condition" that addresses item (i) and a "kinetic relation" that addresses item (ii). In [3] we demonstrate that the solution to the Riemann problem for this quasilinear system is unique when the problem statement is supplemented in this way; see also LeFleeh [34, 35]. Each of the previously mentioned admissibility criteria can be viewed as corresponding to a particular (mathematical) kinetic relation, [4, 5], though deriving the kinetic law from the underlying physics is largely an open question.

1.5 Outline of monograph

This monograph is organized into four informal sections. The first section consists of Chapters 2–4 and introduces the main concepts of the subject within the simplest possible setting, viz. a purely mechanical theory for a one-dimensional elastic bar. The second section, consisting of Chapters 5–8, presents the general three-dimensional theory including both mechanical and thermal effects. In Chapters 9–11, which form the third section, we specialize the general theory to one dimension and then study some specific thermomechanical initial-boundary-value problems. In the final section of the monograph, Chapters 12–14, we consider some three-dimensional problems.

The first and third sections both concern one-dimensional settings; the former excludes thermal effects and the latter includes them. The second and fourth sections both concern three-dimensional settings; the former concerns the general theory and the latter concerns specific problems.

In slightly more detail, Chapter 2 sets out the purely mechanical theory for a one-dimensional elastic bar, Chapter 3 examines static configurations and quasistatic processes for it, and Chapter 4 examines a dynamic problem where inertial effects are important. In Chapter 5, we discuss, within a general three-dimensional context that includes both mechanical and thermal effects, a multiple-well Helmholtz free energy potential; two specific examples, one concerning a van-der Waals fluid and the other a martensitic material, are also presented. Chapter 6 is devoted to developing the general continuum notion of driving force. In Chapter 7 we specialize the results of the preceding chapter to a thermoelastic material. We also examine the stability of such a material. The notions of a kinetic law and nucleation criterion are introduced in Chapter 8, which also includes several examples of phenomenological and micromechanical kinetic relations. Next we specialize the general thermomechanical theory to one dimension, but in contrast to the first section, now we retain thermal effects. In Chapter 9 we describe a constitutive model for two-phase thermoelastic materials of Mie-Grüneisen type. Chapter 10 studies quasistatic thermomechanical hysteresis and the results are compared with experimental observations. Chapter 11 examines a dynamic problem where inertial and thermal effects are both important. Finally we turn to some three-dimensional problems. Chapter 12 concerns statics and elucidates the role of geometric compatibility in determining equilibrium configurations. We turn to dynamics in Chapter 13 and study impact-induced phase transitions in a single crystal; the analysis is used in conjunction with experiments on CuAlNi to determine the kinetic relation for the transformation. In the final chapter, Chapter 14 we examine the quasistatic twinning deformation of a CuAlNi single crystal under biaxial loading.

A different way in which to view the organization of this monograph is as follows:

- chapters devoted to theory (Chapter 2 – one dimension without thermal effects, Chapter 9 – one dimension with thermal effects, and Chapters 5–8 – three dimensions including mechanical and thermal effects);
- chapters devoted to statics (Chapter 3 – one dimension. and Chapter 12 – three dimensions);
- chapters concerned with quasistatic processes (Chapter 3 – one dimension without thermal effects, Chapter 10 – one dimension with thermal effects, and Chapter 14 – three dimensions); and
- chapters devoted to dynamical problems where inertia is important (Chapter 4 – one dimension without thermal effects, Chapter 11 – one dimension with thermal effects, and Chapter 13 – three dimensions).

REFERENCES

[1] R. Abeyaratne, C. Chu, and R.D. James, Kinetics of materials with wiggly energies: theory and application to the evolution of twinning microstructures in a Cu–Al–Ni shape memory alloy. *Philosophical Magazine* A, **73** (1996), pp. 457–97.

[2] R. Abeyaratne and J.K. Knowles, On the driving traction acting on a surface of strain discontinuity in a continuum. *Journal of the Mechanics and Physics of Solids*, **38** (1990), pp. 345–60.

[3] R. Abeyaratne and J.K. Knowles, Kinetic relations and the propagation of phase boundaries in solids. *Archive for Rational Mechanics and Analysis*, **114** (1991), pp. 119–54.

[4] R. Abeyaratne and J.K. Knowles, Implications of viscosity and strain-gradient effects on the propagation of phase boundaries in solids. *SIAM Journal on Applied Mathematics*, **51** (1991), pp. 1205–21.

[5] R. Abeyaratne and J.K. Knowles, On the propagation of maximally dissipative phase boundaries in solids. *Quarterly of Applied Mathematics*, **50** (1992), pp. 149–72.

[6] R Abeyaratne and J.K. Knowles, A note on the driving traction acting on a propagating interface: adiabatic and non-adiabatic processes of a continuum. *ASME Journal of Applied Mechanics*, **67** (2000), pp. 829–31.

[7] R. Abeyaratne and S. Vedantam, A lattice-based model of the kinetics of twin boundary motion. *Journal of the Mechanics and Physics of Solids*, **51** (2003), pp. 1675–700.

[8] M. Achenbach and I. Müller, Simulation of material behavior of alloys with shape memory. *Archive of Mechanics*, **37** (1985), pp. 573–85.

[9] A. Artemev, Y.M. Jin, and A.G. Khachaturyan, Three-dimensional phase field model of proper martensitic transformation. *Acta Materialia*, **49** (2001), pp. 1165–77.

[10] K. Bhattacharya, *Microstructure of Martensite: Why it Forms and How it Gives Rise to the Shape Memory Effect*. Oxford University Press, New York, 2003.

[11] Z.H. Bo and D.C. Lagoudas, Thermomechanical modeling of polycrystalline SMAs under cyclic loading. *International Journal of Engineering Science*, **37** (1999), pp. 1089–140.

[12] M. Brocca, L.C. Brinson, and Z. Bazant, Three-dimensional constitutive model for shape memory alloys based on microplane model. *Journal of the Mechanics and Physics of Solids*, **50** (2002), pp. 1051–77.

[13] K.A. Bywater and J.W. Christian, Martensitic transformations in titanium-tantalum alloys. *Philosophical Magazine*, **25** (1972), pp. 1249–73.

[14] J.W. Christian, *The Theory of Martensitic Transformations in Metals and Alloys*, Part 1. Pergamon, Oxford, 1975.

[15] C. Chu, *Hysteresis and Microstructures: A Study of Biaxial Loading on Compound Twins of Copper–Aluminum-Nickel Single Crystals*. Ph.D. dissertation, University of Minnesota, MN, 1993.

[16] C. Dafermos, Hyperbolic systems of conservation laws, in *Systems of Nonlinear Partial Differential Equations*, edited by J.M. Ball. Reidel, Dordrecht, 1983, pp. 25–70.

[17] C. Dafermos, Discontinuous thermokinetic processes, Appendix 4B of *Rational Thermodynamics*, edited by C. Truesdell. Springer-Verlag, New York, 1984.

[18] T. Duerig, K. Melton, D. Stoeckel, and C.M. Wayman, *Engineering Aspects of Shape Memory Alloys*. Butterworth-Heinemann, London, 1990.

[19] J.L. Ericksen, Equilibrium of bars. *Journal of Elasticity*, **5** (1975), pp. 191–201.

[20] D.J. Erskine and W.J. Nellis, Shock-induced martensitic transformation of highly oriented graphite to diamond. *Journal of Applied Physics*, **71** (1992), pp. 4882–6.

[21] J.C. Escobar and R.J. Clifton, On pressure–shear plate impact for studying the kinetics of stress-induced phase transformations. *Journal of Materials Science and Engineering*, **A170** (1993), pp. 125–42.

[22] J.D. Eshelby, Continuum theory of lattice defects, in *Solid State Physics*, edited by F. Seitz and D. Turnbull, Volume 3. Academic Press, New York, 1956, pp. 79–144.

[23] L.C. Evans, *Partial Differential Equations*. American Mathematical Society, Providence, 1998.

[24] E. Fried and M.E. Gurtin, Dynamic solid–solid transitions with phase characterized by an order parameter. *Physica D*, **72** (1994), pp. 287–308.

[25] H. Funakubo, *Shape Memory Alloys*. Gordon and Breach, New York, 1987.

[26] X. Gao and L.C. Brinson, SMA single crystal experiments and micromechanical modeling for complex thermomechanical loading *Proceedings of SPIE*, **3992** (2000), pp. 516–23.

[27] M.E. Gurtin, *Configuration Forces as Basic Concepts of Continuum Physics*. Springer, New York, 2000.

[28] W. Heidug and F.K. Lehner, Thermodynamics of coherent phase transformations in non-hydrostatically stressed solids. *Pure and Applied Geophysics*, **123** (1985), pp. 91–8.

[29] R.D. James and K.F. Hane, Martensitic transformations and shape-memory materials. *Acta Materialia*, **48** (2000), pp. 197–222.

[30] J.K. Knowles, On the dissipation associated with equilibrium shocks in finite elasticity. *Journal of Elasticity*, **9** (1979), pp. 131–58.

[31] J.K. Knowles and E. Sternberg, On the failure of ellipticity and the emergence of discontinuous deformation gradients in plane finite elastostatics. *Journal of Elasticity*, **8** (1978), pp. 329–79.

[32] R.V. Krishnan and L.C. Brown, Pseudo-elasticity and the strain–memory effect in an Ag-45 at. pct. Cd alloy. *Metallurgical Transactions* A, **4** (1973), pp. 423–9.

[33] P. Lax, *Hyperbolic Systems of Conservation Laws and the Mathematical Theory of Shock Waves*, Regional Conference Series in Applied Mathematics, No. 11. SIAM, Philadelphia, PA, 1973.

[34] P.G. LeFloch, Propagating phase boundaries: formulation of the problem and existence via the Glimm scheme. *Archive for Rational Mechanics and Analysis*, **123** (1993), pp. 153–97.

[35] P.G. LeFloch, *Hyperbolic Systems of Conservation Laws*. Birkhäuser Verlag, Basel, 2002.

[36] P.H. Leo, T.W. Shield, and O. P. Bruno, Transient heat transfer effects on the pseudoelastic behavior of shape memory wires, *Acta Metallurgica et Materialia*, **41** (1993), pp. 2477–85.

[37] T.J. Lim and D.L. McDowell, Mechanical behavior of a NiTi shape memory alloy under axial-torsional proportional and nonproportional loading. *ASME Journal of Engineering Materials and Technology*, **121** (1999), pp. 9–18.

[38] M. Meyers, *Dynamic Behavior of Materials*. Wiley, New York, 1994.

[39] N. Nakanishi, Lattice softening and the origin of SME, in *Shape-Memory Effects in Alloys*, edited by J. Perkins. Plenum Press, New York, 1975, pp. 147–76.

[40] O. Oleinik, On the uniqueness of the generalized solution of the Cauchy problem for a nonlinear system of equations occurring in mechanics. *Uspekhi Matematicheskii Nauk (N.S.)*, **12** (1957), pp. 169–76 (in Russian).

[41] O. Oleinik, Uniqueness and stability of the generalized solution of the Cauchy problem for a quasilinear equation. *Uspekhi Matematicheskii Nauk (N.S.)*, **14** (1959), pp. 165–70 (in Russian).

[42] K. Otsuka and C.M. Wayman, *Shape Memory Materials*. Cambridge University Press, Cambridge, 1998.

[43] R. Pego, Phase transitions in one-dimensional nonlinear viscoelasticity: admissibility and stability. *Archive for Rational Mechanics and Analysis*, **97** (1986), pp. 353–94.

[44] P.K. Purohit, *Dynamics of Phase Transitions in Strings, Beams and Atomic Chains*. Ph.D. Thesis, California Institute of Technology, 2001.

[45] L.M. Schetky, Shape-memory alloys. *Scientific American*, **241** (1979), pp. 74–82.

[46] J.A. Shaw and S. Kyriakides, Thermomechanical aspects of NiTi. *Journal of the Mechanics and Physics of Solids*, **43** (1995), pp. 1243–81.

[47] M. Shearer, Nonuniqueness of admissible solutions of Riemann initial value problems for a system of conservation laws of mixed type. *Archive for Rational Mechanics and Analysis*, **93** (1986), pp. 45–59.

[48] M. Slemrod, Admissibility criteria for propagating phase boundaries in a van der Waals fluid. *Archive for Rational Mechanics and Analysis*, **81** (1983), pp. 301–15.

[49] K. Thornton and P.W. Voorhees, The dynamics of interfaces in elastically stressed solids, in *Thermodynamics, Microstructures and Plasticity*, edited by A. Finel et al. Kluwer, Dordrecht, 2003, pp. 91–106.

[50] L. Truskinovsky, Equilibrium phase interfaces. *Soviet Physics Doklady*, **27** (1982), pp. 551–3.

[51] L. Truskinovsky, Structure of an isothermal phase discontinuity. *Soviet Physics Doklady*, **30** (1985), pp. 945–8.

Part II Purely Mechanical Theory

2 Two-Well Potentials, Governing Equations, and Energetics

2.1 Introduction

In this chapter, we assemble the basic field equations and jump conditions for a one-dimensional, purely mechanical theory of nonlinear elasticity; although thermal effects will be omitted, inertia will be taken into account. The theory presented here is general enough to describe nonlinearly elastic materials that, under suitable conditions of stress, are capable of existing in either of two phases. As we shall see, a key feature of this theory is that the potential energy of the material, as a function of strain at a fixed stress, has two local minima. The associated constitutive relation between stress and strain will then necessarily be nonmonotonic, possessing a maximum and a minimum connected by an unstable regime in which stress declines with increasing strain.

Experiments that provide the motivation for the theory about to be developed fall into two categories. The first of these involves slow tensile loading and unloading of slender bars or wires composed of materials such as shape-memory alloys. The model to be constructed to describe experiments of this kind is one of uniaxial *stress* in a one-dimensional nonlinearly elastic continuum, and the processes to be studied for this model are quasistatic. The stress-induced phase transitions in such experiments occur in tension, so the two minima in the potential energy density occur at positive – or extensional – values of strain, as do the extrema in the stress–strain relation.

In the second type of experiment, the specimen is typically a circular cylindrical disk composed of a metal or a ceramic and subjected to normal impact on one of its plane faces by a projectile or "flyer-plate" moving at high speed. In the model appropriate to these experiments, the disk is replaced by an infinite slab of nonlinearly elastic material whose thickness is that of the disk, and which finds itself in a state of uniaxial *strain* in response to the impact; inertial effects are important. The materials in such experiments undergo phase changes in compression, so the minima in the potential energy and the extrema in the stress–strain relation now occur at negative – or compressive – values of strain.

If inertial effects are retained in both of the nonlinearly elastic models alluded to above, the resulting mathematical theories are identical in the sense that the field equations and jump conditions for both models involve a single scalar strain, a scalar stress constitutively determined by this strain, and a scalar particle velocity. The independent variables are the time and a single space coordinate. For the case of impact in the slab, there are, in addition, transverse components of normal stress that may be of interest in some circumstances, but these do not enter the governing equations and, if needed, are to be calculated after the basic boundary–initial value problem corresponding to a particular experiment has been solved. In general, the state of stress in the slab will *not* be hydrostatic.

For definiteness, we shall develop the one-dimensional theory in this chapter in the context of the thin bar, rather than the slab. We thus deal at first with a general rising–falling–rising stress–strain curve whose maximum and minimum occur at tensile strains, specializing in Chapter 3 to the case of the trilinear material for purposes of detailed analysis of quasistatic motions in the bar. Only minor modifications will prove to be needed to model the slab, for which phase transitions occur in compression. We shall introduce these modifications in Chapter 4, where they will be needed in our analysis of the dynamics of impact-induced phase transitions.

The notion of a two-phase nonlinearly elastic material is introduced in the following section. In Section 2.3, we state the version of global balance of momentum appropriate to the dynamics of uniaxial stress in a one-dimensional bar, and we describe the two-well potential energy density and the associated nonmonotonic nonlinearly elastic stress–strain relation that underlies the model used in this chapter. The balance principle, and the constitutive law, together with the kinematic assumption that displacement is everywhere continuous, provide the fundamental field equations and jump conditions of the theory. Throughout the book, we shall adopt the *Lagrangian*, or *material*, description of the processes considered.

In Section 2.4, we shall study the energetics of motions in the bar, accounting for inertial effects. As in classical gas dynamics, propagating discontinuities in the fields of strain, particle velocity, and stress affect the relation between work and energy in a major way. Indeed, the fact that moving strain discontinuities give rise to an imbalance between the rate of work done on a part of the body containing such jumps and the rate of increase of the total mechanical energy contained in this part leads to the notion of the scalar *driving force acting on a strain discontinuity*. This notion plays a central role in further developments. In particular, the product of the driving force and the referential, or Lagrangian, velocity of the moving discontinuity represents the associated dissipation rate. The requirement that this dissipation rate be nonnegative furnishes the analog of the second law of thermodynamics in the purely mechanical setting of the present chapter.

2.2 Two-phase nonlinearly elastic materials

A nonlinearly elastic material is characterized by an elastic potential $W(\gamma)$ representing *strain energy* per unit reference volume as a function of strain γ. The stress

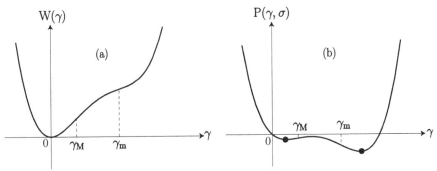

Figure 2.1. (a) The nonconvex strain energy density $W(\gamma)$ for an elastic material that can exist in either of two phases. (b) The associated two-well potential energy density as a function of strain at a fixed stress for which the material can exist in either phase. The function W changes curvature at the strains γ_M and γ_m.

σ is related to the strain γ through W by

$$\sigma = \hat{\sigma}(\gamma) = W'(\gamma), \tag{2.1}$$

and the *potential energy* $P(\gamma, \sigma)$ per unit reference volume is defined as

$$P(\gamma, \sigma) = W(\gamma) - \sigma\gamma. \tag{2.2}$$

Observe that γ and σ are treated as independent variables in P; however at an extremal of P with respect to γ, one has $\partial P/\partial \gamma = 0$, which is equivalent to the constitutive relation $\sigma = \hat{\sigma}(\gamma)$. *Therefore extrema of the potential $P(\cdot, \sigma)$ are of special interest since they correspond to stress, strain pairs related by the constitutive relation.* If the extremal is a *minimum*, or an "energy well," the equilibrium state is stable in a minimum energy sense. Local minima of $P(\cdot, \sigma)$ are identified with material phases, and so, if, for example, P has n energy wells at some stress σ, we say that the material may exist in n phases at that stress.

The three quantities P, W, and σ have certain characteristic properties when associated with a *two-phase material*, that is, a material that can exist in either of two phases for some range of stress. Since the potential energy for such a material must have two local minima for stress in the appropriate range, we shall see that W must be nonconvex and the stress must be a nonmonotonic function of strain.

For the conventional linearly elastic material, (2.1) takes the form $\sigma = \mu\gamma$ where μ is the elastic modulus for uniaxial stress, and so $W(\gamma) = \mu\gamma^2/2$ and $P(\gamma, \sigma) = \mu\gamma^2/2 - \sigma\gamma$. Observe that for any fixed value of σ, this P has a single minimum at $\gamma = \sigma/\mu$.

In general, if W is a convex function of γ then P will have exactly one minimum for any σ. For a *two-phase* material, $P(\cdot, \sigma)$ must have *two* distinct minima for values of stress for which two phases exist. It follows that W cannot be convex for a two-phase material. Figure 2.1(a) shows an example of a nonconvex strain energy function for a two-phase material. Observe that the curvature of W changes at the two strain levels γ_M and γ_m:

$$W''(\gamma_M) = W''(\gamma_m) = 0. \tag{2.3}$$

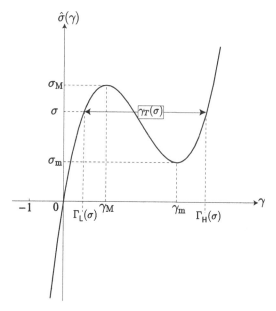

Figure 2.2. The rising–falling–rising stress–strain curve for a two-phase elastic material.

When W is nonconvex as shown in Figure 2.1(a), the associated potential energy P has either one or two energy wells depending on the value of stress. Figure 2.1(b) shows a graph of P for a particular value of σ at which it has two energy wells.

For a two-phase material, $W(\gamma)$ cannot be convex and therefore the stress $\sigma = \hat{\sigma}(\gamma) = W'(\gamma)$ cannot be a monotonically increasing function of strain γ. Figure 2.2 shows a graph of the stress–strain relation $\sigma = \hat{\sigma}(\gamma) = W'(\gamma)$ corresponding to the strain energy function W described in Figure 2.1(a). The strain levels γ_M and γ_m in Figure 2.2 at which the stress attains its local maximum and minimum values respectively are the same strains shown previously in Figure 2.1(a) at which W changes curvature. The corresponding values of stress are $\sigma_M = \hat{\sigma}(\gamma_M)$ and $\sigma_m = \hat{\sigma}(\gamma_m)$. It is readily seen that it is for stress in the interval $[\sigma_m, \sigma_M]$ that the potential $P(\cdot, \sigma)$ has two energy wells. This is precisely the range of stress for which the stress response function $\hat{\sigma}$ is not uniquely invertible.

One can view the classical model of the *van der Waals fluid* as the prototype of continuum-mechanical models of two-phase materials. This is explored further in Section 5.4. The importance of the rising–falling–rising stress–strain curve in modeling phase transitions in solids was made clear in a seminal paper of Ericksen [3].

It is assumed that the stress-response function $\hat{\sigma}(\gamma)$ for the two-phase material characterized by Figure 2.2 is continuous for all $\gamma > -1$, and in class C^2 on $(-1, \gamma_M]$, on $[\gamma_M, \gamma_m]$, and on $[\gamma_m, \infty)$; as we shall see in Section 2.3 the restriction $\gamma > -1$ arises from the requirement that the deformation be one-to-one. Moreover, $\hat{\sigma}(\gamma)$ is to be strictly increasing on $(-1, \gamma_M]$, strictly decreasing on $[\gamma_M, \gamma_m]$, and strictly increasing on $[\gamma_m, \infty)$. Finally, the curvature of the stress–strain curve is assumed to be nonpositive on $(-1, \gamma_M)$, nonnegative on (γ_m, ∞), and to reverse its sign at most once on (γ_M, γ_m). Note that the class of stress-response functions delineated by these requirements includes the so-called *trilinear* elastic material (Figure 2.7) for which each of the three branches of the stress–strain curve is a straight line segment. One of the consequences of these assumptions is the inequality

$$\hat{\sigma}(\gamma_M) > \hat{\sigma}(\gamma_m) \quad \text{or equivalently} \quad \sigma_M > \sigma_m. \tag{2.4}$$

If the strain γ borne by a particle corresponds to a point on the rising branch of the stress–strain curve of Figure 2.2 that includes the origin, the particle is said to be in the *low-strain phase*; strains on the other rising branch of the curve correspond to the *high-strain phase*. The declining branch of the stress–strain curve consists of states that are readily shown to be unstable in the sense that the dynamical equations obtained by linearization about such a state have solutions that grow exponentially in time from smooth initial values representing small departures from the base state.

Although for the stress–strain curve shown in Figure 2.2 one has $\sigma_m > 0$, this need not be the case. If $\sigma_m < 0$, observe from that figure that the high-strain and low-strain phases *both* contain stress-free states. Since P reduces to W when $\sigma = 0$, it follows that W itself must have two energy wells in this case.

As observed already, the material described by Figures 2.1 and 2.2 can exist in one of two phases provided that the stress lies in the interval $\sigma_m \leq \sigma \leq \sigma_M$. The corresponding potential energy $P(\cdot, \sigma)$ has two local minima. Which of these is the global minimum? In order to explore this question it is convenient to express the various physical quantities as functions of stress rather than strain. However, since the stress–strain relation corresponding to Figure 2.2 is not invertible, some care must be taken in doing this. Thus let $\Gamma_L(\sigma)$ be the inverse of the restriction of the stress-response function $\hat{\sigma}(\gamma)$ to the low-strain phase, and similarly let $\Gamma_H(\sigma)$ be the inverse for the high-strain phase. If the stress σ at a particle lies in the interval $[\sigma_m, \sigma_M]$, then excluding unstable states, there are two strains compatible with this stress,

$$\gamma = \Gamma_L(\sigma) \quad \text{and} \quad \gamma = \Gamma_H(\sigma), \qquad \sigma_m \leq \sigma \leq \sigma_M, \tag{2.5}$$

corresponding to the low- and high-strain branches of the stress–strain curve; see Figure 2.2. The *transformation strain* $\gamma_T(\sigma)$ is the strain jump suffered by a particle in transforming from the low-strain phase to the high-strain phase at the stress σ:

$$\gamma_T(\sigma) = \Gamma_H(\sigma) - \Gamma_L(\sigma), \quad \sigma_m \leq \sigma \leq \sigma_M. \tag{2.6}$$

Let

$$g_L(\sigma) = P(\Gamma_L(\sigma), \sigma), \quad \sigma_{-1} < \sigma \leq \sigma_M, \qquad g_H(\sigma) = P(\Gamma_H(\sigma), \sigma), \quad \sigma \geq \sigma_m, \tag{2.7}$$

be the values of the potential energy $P(\cdot, \sigma)$ at its two minima; in (2.7), we have set $\sigma_{-1} = \hat{\sigma}(-1+)$, which may take the value $-\infty$. The functions g_L and g_H are the counterparts in the present purely mechanical setting of the *Gibbs free energies* of the two material phases. Let

$$e(\sigma) = g_L(\sigma) - g_H(\sigma), \quad \sigma_m \leq \sigma \leq \sigma_M, \tag{2.8}$$

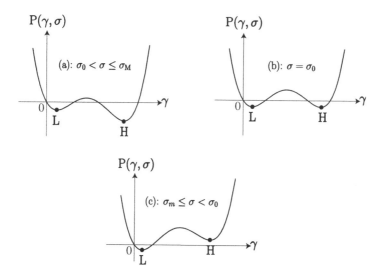

Figure 2.3. Potential energy for a two-phase material as a function of strain for a fixed value of stress (a) above, (b) at, and (c) below the Maxwell stress σ_0. In each case, the coordinates of the point L corresponding to the low-strain energy well are $(\Gamma_L(\sigma), g_L(\sigma))$, while those of H are $(\Gamma_H(\sigma), g_H(\sigma))$; see (2.5) and (2.7).

be the difference between these two Gibbs energies at the same stress σ. One can show that the derivative $e'(\sigma)$ is given by

$$e'(\sigma) = \gamma_T(\sigma) > 0, \tag{2.9}$$

so that $e(\sigma)$ is an increasing function of σ on $[\sigma_m, \sigma_M]$. Moreover $e(\sigma_m) < 0$, while $e(\sigma_M) > 0$. There is therefore a unique value σ_0 of stress – the *Maxwell stress* – at which the values of the two minima of P, and so the two Gibbs energies, coincide:

$$g_L(\sigma_0) = g_H(\sigma_0). \tag{2.10}$$

For $\sigma_m \leq \sigma < \sigma_0$, the low-strain minimum of $P(\cdot, \sigma)$ lies below the high-strain minimum, while for $\sigma_0 < \sigma \leq \sigma_M$, the reverse is true. We thus speak of the low-strain phase as *stable* and the high-strain phase as *metastable* below the Maxwell stress and vice versa above the Maxwell stress. *At* the Maxwell stress, both phases are stable. Figure 2.3 shows schematic plots of $P(\gamma, \sigma)$ versus γ for values of σ above, at, and below the Maxwell stress for a material whose stress–strain curve is as shown in Figure 2.2.

It is easy to show that (2.10) can be rewritten, after using (2.7), (2.2), and (2.1), as

$$\int_{\Gamma_L(\sigma_0)}^{\Gamma_H(\sigma_0)} \hat{\sigma}(\gamma) \, d\gamma = \sigma_0 \left(\Gamma_H(\sigma_0) - \Gamma_L(\sigma_0) \right), \tag{2.11}$$

which is an alternative equation for the Maxwell stress σ_0. This shows that the Maxwell stress cuts lobes of equal area from the rising–falling–rising stress–strain curve of Figure 2.2.

2.3 Field equations and jump conditions

Consider a one-dimensional bar that occupies the interval $0 \leq x \leq L$ of the x-axis in a stress-free reference configuration. During a motion of this body, the particle located at x in the reference configuration is carried at time t to the point $y = x + u(x, t)$ where $u(x, t)$ denotes the displacement. It is assumed throughout this book that the displacement is continuous, with piecewise continuous first and second derivatives with respect to both spatial and time coordinates. The strain γ and the particle velocity v are defined by $\gamma = u_x$ and $v = u_t$, where subscripts indicate partial derivatives. Observe that the degree of smoothness assumed allows γ and v to have jump discontinuities; this is essential for modeling deformations that involve *co-existent* phases.

The deformation $y = x + u(x, t)$ will be one-to-one if y is a monotonically increasing function of x, that is, if $u_x > -1$; as in the preceding section, we therefore require that $\gamma = u_x > -1$.

We next set out the basic field equations and jump conditions; for more detailed discussion, the reader is referred, for example, to the book of Chadwick [2]. Let $\sigma(x, t)$ denote the nominal longitudinal stress in the bar. Consider a portion of the bar that lies between $x = x_1$ and $x = x_2$ in the reference state; see Figure 2.4. In the absence of body force, balance of momentum of the sub-bar $[x_1, x_2]$ requires that

$$\sigma(x_2, t) - \sigma(x_1, t) = \frac{d}{dt} \int_{x_1}^{x_2} \rho v(x, t) dx. \qquad (2.12)$$

We shall take ρ, the mass per unit referential volume, to be independent of x. The global balance law (2.12) must hold for all x_1 and x_2 ($> x_1$) in $[0, L]$.

If the fields γ, σ, and v are smooth between x_1 and x_2, then the balance law for momentum (2.12) implies the field equation

$$\sigma_x = \rho v_t. \qquad (2.13)$$

Further, smoothness of $\gamma = u_x$ and $v = u_t$ requires that

$$v_x = \gamma_t. \qquad (2.14)$$

Next, suppose that the motion involves a single strain discontinuity, whose referential location at time t is $x = s(t)$. Suppose that this discontinuity lies between x_1 and x_2. Then the global balance of momentum (2.12) and the assumed continuity

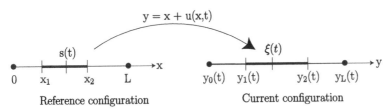

Figure 2.4. Reference and current configurations of a bar.

of displacement require the jump conditions

$$\llbracket \sigma \rrbracket = -\rho \dot{s} \llbracket v \rrbracket, \tag{2.15}$$

$$\llbracket v \rrbracket = -\dot{s} \llbracket \gamma \rrbracket \tag{2.16}$$

to hold at $x = s(t)$. Here we are using the notation

$$\llbracket \phi \rrbracket = \phi^+ - \phi^- = \phi(s(t)+, t) - \phi(s(t)-, t) \tag{2.17}$$

to denote the jump in any field quantity $\phi(x, t)$ at $x = s(t)$. Note that $\dot{s}$ is the referential (or *Lagrangian*) velocity of the discontinuity. Since the location of the discontinuity at time t in the laboratory frame is $\xi(t) = s(t) + u(s(t), t)$, its laboratory (or *Eulerian*) velocity is given by

$$\dot{\xi}(t) = \dot{s}(t) + \gamma^-(t)\dot{s}(t) + v^-(t) = \dot{s}(t) + \gamma^+(t)\dot{s}(t) + v^+(t). \tag{2.18}$$

That the two representations of $\dot{\xi}(t)$ in (2.18) are equivalent is assured by (2.16).

The bar is in a state of *mechanical equilibrium* if u is independent of time; in this event, γ is also time independent, and v vanishes identically. It then follows from (2.13) and (2.15) that the stress σ is independent of x throughout the bar.

The two partial differential equations (2.13) and (2.14), together with the constitutive law

$$\sigma = \hat{\sigma}(\gamma), \tag{2.19}$$

provide three field equations for the three unknowns σ, γ, and v. The associated jump conditions are given by (2.15) and (2.16). The field equations reduce immediately to the following pair of quasilinear partial differential equations:

$$\left. \begin{array}{l} \hat{\sigma}'(\gamma)\gamma_x = \rho v_t, \\[2mm] v_x = \gamma_t; \end{array} \right\} \tag{2.20}$$

these in turn can be reduced further to the single quasilinear equation

$$\left[\hat{\sigma}'(\gamma)\gamma_x \right]_x = \rho \gamma_{tt} . \tag{2.21}$$

The type of the partial differential equation (2.21) is determined by the sign of the slope $\hat{\sigma}'(\gamma)$ of the stress–strain curve. At a solution γ and at a point x and time t, the partial differential equation (2.21) is hyperbolic if $\gamma(x, t)$ is in either the low-strain phase or the high-strain phase (i.e. $\hat{\sigma}'(\gamma) > 0$) and elliptic if $\gamma(x, t)$ corresponds to an unstable state on the declining branch of the stress–strain curve (i.e. $\hat{\sigma}'(\gamma) < 0$).

Suppose $\gamma(x, t)$ suffers a jump discontinuity at $x = s(t)$. Eliminating v from the jump conditions (2.15) and (2.16) yields the following relation between the velocity $\dot{s}$ of the discontinuity and the strains $\gamma^\pm$ on either side:

$$\rho \dot{s}^2 = \frac{\hat{\sigma}(\gamma^+) - \hat{\sigma}(\gamma^-)}{\gamma^+ - \gamma^-}. \tag{2.22}$$

According to (2.22), a discontinuity can propagate with real speed $\dot{s}$ if and only if the chord connecting the two points $(\gamma^{\pm}, \hat{\sigma}(\gamma^{\pm}))$ on the stress–strain curve has nonnegative slope. In particular, no such discontinuity can correspond to two points on the unstable branch of the curve. We call the discontinuity a *shock wave* if the strains $\gamma^{\pm}$ are either both in the low-strain phase or both in the high-strain phase. If, on the other hand, γ^{+} is in the low-strain phase while γ^{-} is in the high-strain phase, or vice versa, the discontinuity is called a *phase boundary*. It will become clear in the next section that we need not consider discontinuities in which one of the strains $\gamma^{\pm}$ lies on the unstable branch of the stress–strain curve.

2.4 Energetics of motion, driving force, and dissipation inequality

Consider again the portion of the bar consisting of particles whose positions in the reference configuration lie between x_1 and x_2. Let

$$D(t) = \sigma v \Big|_{x_1}^{x_2} - \frac{d}{dt} \int_{x_1}^{x_2} \left(\frac{1}{2} \rho v^2 + W(\gamma) \right) dx \qquad (2.23)$$

be the *dissipation rate*, that is, the difference between the rate of work of the external forces acting on the sub-bar $[x_1, x_2]$ and the rate of change of total mechanical energy within it, both measured per unit cross-sectional area. Suppose there is exactly one moving strain discontinuity in $[x_1, x_2]$, and let $x = s(t)$ be its referential location at time t.

Making use of the field equations and jump conditions, one can show that

$$D(t) = f(t)\dot{s}(t), \qquad (2.24)$$

where $f(t)$ is called the *driving force acting on the strain discontinuity* and is given by $f(t) = \hat{f}(\gamma^{+}(t), \gamma^{-}(t))$. Here $\gamma^{\pm} = \gamma^{\pm}(t) = \gamma(s(t)\pm, t)$ are the strains immediately ahead of and behind the strain discontinuity and the function $\hat{f}$ is defined, for $\gamma^{-} > -1$, $\gamma^{+} > -1$, by

$$\begin{aligned} f = \hat{f}(\gamma^{+}, \gamma^{-}) &= [\![W(\gamma)]\!] - (1/2)\left(\hat{\sigma}(\gamma^{+}) + \hat{\sigma}(\gamma^{-}) \right) [\![\gamma]\!] \\ &= \int_{\gamma^{-}}^{\gamma^{+}} \hat{\sigma}(\gamma) d\gamma - (1/2)(\hat{\sigma}(\gamma^{+}) + \hat{\sigma}(\gamma^{-}))(\gamma^{+} - \gamma^{-}). \end{aligned} \qquad (2.25)$$

Since $\hat{f}(\gamma^{+}, \gamma^{-})$ is defined for *all* strains $\gamma^{\pm}$, the stresses $\sigma^{\pm} = \hat{\sigma}(\gamma^{\pm})$ need not lie in $[\sigma_m, \sigma_M]$.

Equations (2.23)–(2.25) apply at a strain discontinuity in *any* nonlinearly elastic material. For a two-phase material, there are two kinds of strain discontinuities, viz. shock waves and phase boundaries, and the notion of driving force and its relation to the dissipation rate are relevant at *both*.

The concept of driving force, if not the name, was independently employed in the setting of phase transitions by Heidug and Lehner [5], Truskinovsky [9], and Abeyaratne and Knowles [1]. It is related to notions used earlier by Eshelby [4] and

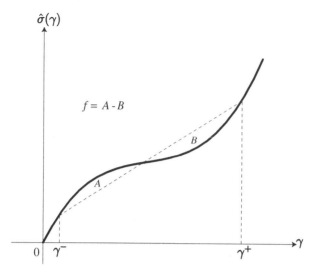

Figure 2.5. Geometric interpretation of the driving force $f = \hat{f}(\gamma^+, \gamma^-)$ at a strain discontinuity as the difference of the signed areas A and B.

Rice [8] in more general settings, and by Knowles [6] for quasistatic processes in nonlinearly elastic solids. There is also a close connection to the idea of driving force that plays a prominent role in materials science; see, for example, Section 1.9 of [7].

In our study of mixed-phase equilibria in the next chapter, we shall see that the *static* version of driving force is directly related to the energy difference $e(\sigma)$ introduced in (2.8).

For any $\hat{\sigma}(\gamma)$, one may interpret $f = \hat{f}(\gamma^+, \gamma^-)$ given by $(2.25)_3$ geometrically as the (signed) difference between the area under the stress–strain curve between γ^- and γ^+ and the area of the trapezoid formed by the following four points in the γ, σ-plane: $(\gamma^-, 0)$, $(\gamma^-, \hat{\sigma}(\gamma^-))$, $(\gamma^+, 0)$ and $(\gamma^+, \hat{\sigma}(\gamma^+))$; see Figure 2.5.

It follows as an immediate consequence of this interpretation that if $\hat{\sigma}(\gamma)$ is linear, then the driving force vanishes at every strain discontinuity; all such discontinuities are therefore dissipation free for a linearly elastic material.

If there is in fact *no* discontinuity at $x = s(t)$, so that $[\![\gamma]\!] = 0$, then it follows from (2.25) and (2.24) that $f = 0$ and $D = 0$; hence the rate of work coincides with the rate of change of total energy, and the rate of dissipation vanishes. This is the usual setting of nonlinear elasticity, which is dissipation free only because moving strain discontinuities are absent. When a genuine discontinuity *is* present, the driving force – and therefore D – need not vanish, though they *may* do so in the exceptional case for which the states $(\gamma^-, \hat{\sigma}(\gamma^-))$ and $(\gamma^+, \hat{\sigma}(\gamma^+))$ are such that the areas A and B in Figure 2.5 happen to be equal. Thus moving strain discontinuities in general produce dissipation in nonlinear elasticity theory, just as shocks waves do when modeled as in classical inviscid gas dynamics. In the present purely mechanical setting, the counterpart of the second law of thermodynamics is the requirement that, at any strain discontinuity, the dissipation rate $D(t)$ must be nonnegative. Thus the driving force f and the Lagrangian velocity $\dot{s}$ of the

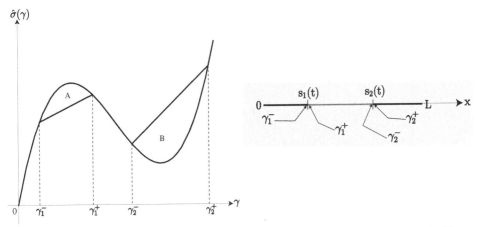

Figure 2.6. The signed areas A and B corresponding to the driving forces acting on discontinuities at the ends of an interval in which the strain lies in the unstable regime of the stress–strain curve.

discontinuity must satisfy the *dissipation inequality*:

$$f(t)\dot{s}(t) \geq 0. \tag{2.26}$$

The dissipation inequality has far-reaching consequences. Consider a two-phase elastic material and suppose, for example, that $\gamma(x, t)$ is a solution of (2.21) for which, during some time interval, one has $\gamma_M < \gamma(x, t) < \gamma_m$ for $s_1(t) < x < s_2(t)$, corresponding to states on the dynamically unstable, declining branch of the stress–strain curve. Suppose further that $\gamma(x, t)$ has jump discontinuities at $x = s_1(t)$ and at $x = s_2(t)$, and let $\gamma_1^{\pm}(t)$ and $\gamma_2^{\pm}(t)$ be the respective strain pairs on either side of these discontinuities. First we note that the relation (2.22) prohibits a discontinuity for which the strains on either side both lie in the unstable range, since the speed of propagation of such a jump would not be real. Thus $\gamma_1^-(t)$ and $\gamma_2^+(t)$ must each correspond to a point on a rising branch of the stress–strain curve. Suppose, for example, that γ_1^- is in the low-strain phase, while γ_2^+ is in the high-strain phase; see Figure 2.6. Under the assumptions made above concerning $\hat{\sigma}(\gamma)$ and as shown in the figure, the chord connecting the points $(\gamma_1^-, \hat{\sigma}(\gamma_1^-))$ and $(\gamma_1^+, \hat{\sigma}(\gamma_1^+))$ on the stress–strain curve lies below the curve itself. Moreover, $\gamma_1^- < \gamma_1^+$. It follows from $(2.25)_2$ that the signed area A in Figure 2.6 is positive, so that the driving force $f = \hat{f}(\gamma_1^+, \gamma_1^-)$ acting on the strain discontinuity at $x = s_1(t)$ is positive as well. The dissipation inequality (2.26) then implies that $\dot{s}_1(t) \geq 0$. On the other hand, the signed area B in Figure 2.6 is negative, which, together with (2.26), shows that $\dot{s}_2(t) \leq 0$. Thus the length $s_2(t) - s_1(t)$ of the interval bearing strains in the unstable regime cannot increase with time t. One obtains the same conclusion regardless of the rising branch of the stress–strain curve that contains γ_1^- or γ_2^+. We have therefore established the following result:

If the values $\gamma(x, t)$ of a solution of (2.21) lie on the unstable branch of the stress–strain curve for x in a time-dependent interval that terminates in jump discontinuities of γ, then the length of this interval cannot increase with time.

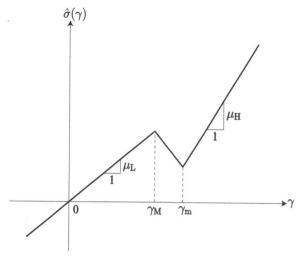

Figure 2.7. The general trilinear two-phase elastic material.

It follows that, if there is an instant t_0 at which there is no x-interval bearing only strains in the unstable interval (γ_M, γ_m) and terminating in a strain discontinuity at each end, then no such interval can arise for $t > t_0$. In the boundary–initial value problems with which we shall be concerned below, the initial data will never contain strains in the unstable regime, so that intervals bearing such strains *and terminating in discontinuities* cannot arise in the solutions.

To see another illustration of the application of the dissipation inequality, consider for a moment the subclass of stress-response functions for which $\hat{\sigma}''(\gamma)$ is strictly negative in the low-strain phase and strictly positive in the high-strain phase. In such a material, the dissipation inequality restricts the nature of shock waves in the following sense. Since the chord connecting two points in the low-strain phase on the stress–strain curve lies below the curve itself, the sign of the driving force at a low-strain phase shock wave coincides with the sign of $\gamma^+ - \gamma^-$. It then follows from (2.26) that $(\gamma^+ - \gamma^-)\dot{s} \geq 0$. Thus a low-strain phase shock wave moves into the regime of greater strain. In the high-strain phase, the reverse is true. In particular, if a low-strain-phase shock moves into undisturbed material ($\dot{s} > 0$, $\gamma^+ = v^+ = 0$), the strain behind the shock must be contractional, that is, $\gamma^- < 0$; in this case, (2.18) shows that the Lagrangian and Eulerian shock speeds coincide: $\dot{\xi} = \dot{s}$.

There is a special case of the stress-response function represented by Figure 2.2 for which the associated driving force of (2.25) has especially simple properties and that plays a major role in the detailed discussions in the following two chapters. For the class of so-called *trilinear* two-phase elastic materials, each of the three branches of the stress–strain curve is a straight line segment, as in Figure 2.7. For this material, the geometric interpretation of the driving force immediately reveals that, for a shock wave in either the low- or the high-strain phase, the driving force acting on the discontinuity vanishes. It then follows that the dissipation inequality (2.26) is trivially satisfied by such a shock wave. Thus restrictions of the kind arising when the stress–strain curve is strictly concave in the low-strain phase and strictly convex in the high-strain phase do not arise for the trilinear material. Moreover, this material is simpler than the more general one of Figure 2.2 in that the respective speeds c_L and c_H of shock waves in the low- and high-strain phases are, from (2.22), known independently of the local states of strain at the shock. They are given by

$$c_L = \sqrt{\mu_L/\rho}, \qquad c_H = \sqrt{\mu_H/\rho}, \tag{2.27}$$

where μ_L and μ_H are the constant elastic moduli of the two phases, shown in the figure as the slopes of the two rising branches. Trilinear materials will be examined in more detail in the next two chapters.

REFERENCES

[1] R. Abeyaratne and J.K. Knowles, On the driving traction acting on a surface of strain discontinuity in a continuum. *Journal of the Mechanics and Physics of Solids*, **38** (1990), pp. 345–60.

[2] P. Chadwick, *Continuum Mechanics*. Dover, New York, 1999.

[3] J.L. Ericksen, Equilibrium of bars. *Journal of Elasticity*, **5** (1975), pp. 191–201.

[4] J.D. Eshelby, Continuum theory of lattice defects, in *Solid State Physics*, edited by F. Seitz and D. Turnbull, Volume 3. Academic Press, New York, 1956, pp. 79–144.

[5] W. Heidug and F.K. Lehner, Thermodynamics of coherent phase transformations in non-hydrostatically stressed solids. *Pure and Applied Geophysics*, **123** (1985), pp. 91–8.

[6] J.K. Knowles, On the dissipation associated with equilibrium shocks in finite elasticity. *Journal of Elasticity*, **9** (1979), pp. 131–58.

[7] D.A. Porter and K.E. Easterling, *Phase Transformations in Metals and Alloys*. Van Nostrand-Reinhold (UK) Ltd, London, 1981.

[8] J.R. Rice, Continuum mechanics and thermodynamics of plasticity in relation to microscale deformation mechanisms, in *Constitutive Equations in Plasticity*, edited by A. S. Argon. MIT Press, Cambridge, MA, 1975, pp. 23–79.

[9] L. Truskinovsky, Structure of an isothermal discontinuity. *Soviet Physics Doklady*, **30** (1985), pp. 945–8.

3 Equilibrium Phase Mixtures and Quasistatic Processes

3.1 Introduction

In the present chapter we study the equilibrium and quasistatic response of a thin bar composed of a material modeled by the stress–strain curve shown in Figure 2.2 of the preceding chapter. There is an extensive experimental literature devoted to tensile loading and unloading of bars made of materials that are capable of undergoing displacive phase transitions; often the materials studied are technologically important shape-memory alloys such as nickel–titanium. For a small sample of this literature, the reader might consult the papers of Krishnan and Brown [13], Nakanishi [17], Shaw and Kyriakides [19], and Lin et al. [14], as well as the references cited there. The loading in such experiments is slow, in the sense that inertia is insignificant. The objective is typically the determination of the relation between the applied stress and the overall elongation of the bar, though in some studies, such as that of Shaw and Kyriakides [19], local strain and temperature measurements are made as well. The stress–elongation relation that is observed in such experiments exhibits hysteresis, the phase transition being the primary mechanism responsible for such dissipative behavior. For a given material, the size and other qualitative features of the hysteresis loops depend on the loading rate and the temperature at which the test takes place. In the model to be discussed in this chapter, thermal effects are omitted; they will be accounted for in later chapters.

Our interest here is twofold: first, the purely mechanical equilibrium theory provides the simplest setting in which to show that, if unconventional material models such as that represented by Figure 2.2 are to be studied, the standard problems of continuum mechanics are in fact indeterminate when posed in terms of the usual balance principles, boundary conditions, and the bulk constitutive law. If such indeterminacy is to be removed, additional information must be supplied in formulating the basic problems. Second, we wish to assess the extent to which the purely mechanical theory of one-dimensional bars composed of two-phase nonlinearly elastic materials can predict the experimentally observed features of the quasistatic hysteresis loops.

We shall study a one-dimensional tensile bar composed of the material represented by Figure 2.2. In the next section, we examine equilibrium mixtures of the two phases of this material. The variational theory of such mixed-phase equilibria is described in Section 3.3; the determination of two-phase minimum-energy equilibria provides an alternative route to the concept of driving force, a concept that was introduced in Chapter 2 in connection with dissipation in dynamic processes. In Section 3.4, we present the general theory of one-dimensional quasistatic processes in a two-phase elastic material; these are treated as one-parameter families of equilibrium states in which the parameter is time. We show that the response of the bar to a given loading history, even with the dissipation inequality enforced, is in general indeterminate. In Section 3.5 we describe continuum-mechanical versions of the notions of a *nucleation criterion* and a *kinetic relation* that are familiar from materials science. The theory, enhanced with these two additional ingredients, is applied in Section 3.6 to constant-elongation rate loading–unloading processes of the kind commonly found in experiments. With nucleation and kinetics included, the model for quasistatic processes predicts a hysteretic relation between stress and the overall elongation of the bar that exhibits many of the qualitative features of the hysteresis loops observed in experiments. We discuss this hysteresis in Section 3.7.

3.2 Equilibrium states

A tensile bar, which occupies the interval $0 \leq x \leq L$ of the x-axis in an unstressed reference configuration, is deformed to a state of equilibrium. Body forces are absent. The particle at x in the undeformed state is carried to $x + u(x)$ by the deformation, where u is the axial displacement. The end $x = 0$ of the bar is taken to be fixed, so that $u(0) = 0$. The strain associated with this deformation is $\gamma(x) = u'(x)$, where prime indicates derivative. The constitutive law for the nonlinearly elastic material of the bar is that of equation (2.1) and the associated stress–strain curve is that of Figure 2.2. In particular, recall that $\gamma = \Gamma_{\mathrm{L}}(\sigma)$ and $\gamma = \Gamma_{\mathrm{H}}(\sigma)$ describe the stress–strain behavior of the low- and high-strain phases; they are the respective inverses of the restrictions of the stress response function $\hat{\sigma}$ to each phase.

In the absence of inertial effects, equations (2.13) and (2.15) show that the stress σ must be uniform throughout the bar.

Suppose first that the stress is not too large; specifically, that it lies in the range $\sigma_{-1} < \sigma < \sigma_m$ where σ_m is the stress at the local minimum in Figure 2.2, and we have set $\sigma_{-1} = \hat{\sigma}(-1+)$. For this range of stress it can be seen from Figure 2.2 that every particle of the bar must be in the low-strain phase. The strain field is therefore uniform and is given by $\gamma(x) = \Gamma_{\mathrm{L}}(\sigma)$, $0 \leq x \leq L$. Similarly for large values of stress, specifically for $\sigma > \sigma_M$, the bar is necessarily in the high-strain phase, the strain in the bar is uniform, and it is given by $\gamma(x) = \Gamma_{\mathrm{H}}(\sigma)$, $0 \leq x \leq L$.

Next suppose that the stress σ has a value in the intermediate range $\sigma_m \leq \sigma \leq \sigma_M$. In this case a particle can be in either the low-strain phase or the high-strain phase with corresponding strains $\Gamma_{\mathrm{L}}(\sigma)$ or $\Gamma_{\mathrm{H}}(\sigma)$. Therefore for this range of stress, an equilibrium configuration can involve a *mixture of coexistent phases* with some particles being in the low-strain phase and others in the high-strain phase. For definiteness, consider a configuration in which particles in the range $0 \leq x < s$ lie in the high-strain phase while particles in the range $s < x \leq L$ lie in the low-strain phase for some $s \in [0, L]$. The corresponding piecewise homogeneous strain distribution in the bar is given by

$$\gamma(x) = \begin{cases} \Gamma_H(\sigma), & 0 \leq x < s, \\ \Gamma_L(\sigma), & s < x \leq L. \end{cases} \tag{3.1}$$

The strain field (3.1) involves a *phase boundary* at $x = s$, the high-strain phase being on its left and the low-strain phase on its right. When $s = 0$, the bar is entirely in the low-strain phase, while for $s = L$, it is in the high-strain phase. More precisely, $x = s$ is the *referential location* of the phase boundary; an observer viewing the deformed bar in the laboratory sees the phase boundary at the point $s + u(s) = s(1 + \Gamma_H(\sigma))$ on the x-axis.

Since the displacement field is to be continuous, we have

$$u(x) = \begin{cases} \Gamma_H(\sigma)x, & 0 \leq x \leq s, \\ \Gamma_L(\sigma)x + \gamma_T(\sigma)s, & s \leq x \leq L, \end{cases} \tag{3.2}$$

where $\gamma_T(\sigma) = \Gamma_H(\sigma) - \Gamma_L(\sigma)$ is the transformation strain. The overall elongation $u(L) = \delta$ of the bar is given by

$$\delta = \hat{\delta}(\sigma, s) = \Gamma_H(\sigma)s + \Gamma_L(\sigma)(L - s). \tag{3.3}$$

For a given stress $\sigma \in [\sigma_m, \sigma_M]$, equation (3.3) determines a one-parameter family of elongations $\delta = \hat{\delta}(\sigma, s)$. For any fixed s in the interval $[0, L]$, the right side of (3.3) is a strictly increasing function of σ, so that for a given elongation, (3.3) determines a one-parameter family of stresses $\sigma = \bar{\sigma}(\delta, s)$.

The following important observation may be made concerning the relation (3.3) among δ, σ, and s. If $u(L) = \delta$ is regarded as prescribed, one might wish to find the stress σ, thus solving the simplest of boundary value problems in nonlinear elasticity: the displacement problem in a one-dimensional tensile bar. But clearly (3.3) fails to determine σ, because of the presence of the unknown phase boundary location s. To render the solution of this elementary problem unique, one must supply additional information beyond the balance of momentum, the constitutive law, the boundary conditions, and the smoothness specifications. Ordinarily, these requirements are sufficient, in the absence of instabilities such as buckling, to assure uniqueness for conventional elastic materials for which stress is a monotonically increasing function of strain.

The same lack of uniqueness arises from (3.3) if σ, rather than δ, is prescribed. Thus for two-phase elastic materials, uniqueness is lost for both the *displacement*- and *stress*-boundary value problems in the equilibrium theory of the tensile bar. This uniqueness breakdown must be resolved by supplying a mechanism for the determination of the "internal variable" s.

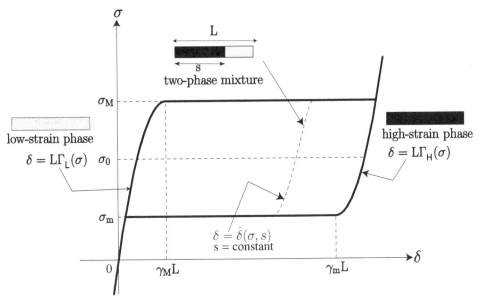

Figure 3.1. The "flag" of mixed-phase equilibrium states arising from (3.1) and (3.3), together with the "flagpoles" of pure low-strain and high-strain equilibria and a generic curve of constant s.

The collection of equilibrium phase mixtures described by (3.1) and (3.3) may be represented by the set of all pairs (δ, σ) obtained from (3.3) as σ varies between σ_m and σ_M and s ranges over the interval $[0, L]$. This set is shown in the δ, σ-plane as Figure 3.1. The figure includes the stress-elongation curves $\delta = \Gamma_L(\sigma)L$ and $\delta = \Gamma_H(\sigma)L$ that describe the respective single-phase responses of the bar in the low- and high-strain phases. We speak of the set described in the figure as the "flag" of mixed-phase equilibria, together with the two "flagpoles" corresponding to the pure phases.

In seeking to understand and resolve the indeterminacy of the displacement boundary value problem, we shall find that the driving force plays a significant role.

Since σ is uniform in the equilibrium phase mixture, the general representation (2.25) for the driving force in terms of the strains $\gamma^{\pm}$ specializes to

$$f = \int_{\gamma^-}^{\gamma^+} \hat{\sigma}(\gamma)\, d\gamma - \sigma(\gamma^+ - \gamma^-) \qquad (3.4)$$

in the static case. Geometrically, the static driving force coincides with the difference of the areas of the two lobes of the stress–strain curve (Figure 2.2) cut off by the line parallel to the γ-axis through the stress σ. Clearly, one may rewrite (3.4) as

$$f = [\![W - \sigma\gamma]\!] = [\![P(\gamma, \sigma)]\!], \qquad (3.5)$$

where $P(\gamma, \sigma)$ is the potential energy per unit reference volume. Thus *for an equilibrium mixture of phases, the driving force acting on the phase boundary at $x = s$ coincides with the jump in the potential energy density across this interface.* Going

further, one notes from (3.5), (2.7), and (2.8) that the static driving force also coincides with the difference $e(\sigma)$ of the minimum values of potential energy at the
stress σ:

$$f = \mathsf{f}(\sigma) = e(\sigma) = \int_{\sigma_0}^{\sigma} \gamma_T(\sigma')\, d\sigma', \tag{3.6}$$

where σ_0 is the Maxwell stress and $\gamma_T(\sigma)$ is the transformation strain at the stress
σ; see (2.6). We see, incidentally, that the Maxwell stress may be interpreted geometrically as the stress at which the two lobes cut from the stress–strain curve as
mentioned above have equal area. Since the stress in a mixed-phase equilibrium
state must lie between σ_m and σ_M, the range of possible values of driving force is
$[f_m, f_M]$, where, by (3.6),

$$f_m = -\int_{\sigma_m}^{\sigma_0} \gamma_T(\sigma)\, d\sigma < 0, \qquad f_M = \int_{\sigma_0}^{\sigma_M} \gamma_T(\sigma)\, d\sigma > 0. \tag{3.7}$$

From (3.6), it follows that the driving force acting on the interface between
the phases vanishes if the stress in the bar coincides with the Maxwell stress. In
this case, we say the bar is in *phase equilibrium*. Note that the bar may sustain a
mixture of phases that is in *mechanical equilibrium* (σ is uniform) but not in phase
equilibrium ($\sigma \neq \sigma_0$).

The total energy E per unit cross-sectional area of the bar due to the strain field
(3.1) is given by

$$E = \int_0^L W(\gamma(x))\, dx = W(\Gamma_{\mathsf{H}}(\sigma))s + W(\Gamma_{\mathsf{L}}(\sigma))(L - s). \tag{3.8}$$

In view of the remarks following (3.3), we may use $\sigma = \bar{\sigma}(\delta, s)$ to regard E as a
function of the macroscopic variable δ and the *internal variable* s: $E = \bar{E}(\delta, s)$. By
direct calculation, one can show that

$$\frac{\partial \bar{E}(\delta, s)}{\partial \delta} = \sigma, \qquad \frac{\partial \bar{E}(\delta, s)}{\partial s} = -f. \tag{3.9}$$

The total potential energy Π is

$$\Pi = \int_0^L P(\gamma(x), \sigma)\, dx = E - \sigma\delta. \tag{3.10}$$

Clearly, we may regard Π as a function of σ and s: $\Pi = \hat{\Pi}(\sigma, s)$; one then shows
that

$$\frac{\partial \hat{\Pi}(\sigma, s)}{\partial \sigma} = -\delta, \qquad \frac{\partial \hat{\Pi}(\sigma, s)}{\partial s} = -f. \tag{3.11}$$

Thus the pairs (δ, σ) and $(s, -f)$ are conjugates with respect to $\bar{E}$, and $(\sigma, -\delta)$ and
$(s, -f)$ are conjugate pairs with respect to $\hat{\Pi}$.

3.3 Variational theory of equilibrium mixtures of phases

We next turn to the variational description of the equilibrium theory of phase mixtures. An illuminating discussion of this subject may be found in the important paper of Ericksen [7].

We begin by asking the following question: among all the mixed-phase equilibria described by (3.1) and (3.3), which one is "preferred" – or "stable" – in the sense that it minimizes the associated total energy $\bar{E}(\delta, s)$ for a given fixed δ, as s varies over $[0, L]$? After answering this question, we shall generalize it by greatly enlarging the class of candidate displacement fields over which to seek the minimum of the total energy.

From the relations (3.9), one has

$$\frac{\partial \bar{E}(\delta, s)}{\partial s} = -f = -\mathsf{f}(\sigma) = -\mathsf{f}(\bar{\sigma}(\delta, s)), \tag{3.12}$$

where the driving force $\mathsf{f}(\sigma)$ is given by (3.6). One observes from (3.12) that, for a given δ, a necessary condition for the total energy $\bar{E}(\delta, s)$ to have an extremum is that s be chosen so that $\bar{\sigma}(\delta, s) = \sigma_0$; then from (3.3),

$$s = \frac{\delta - \Gamma_{\mathsf{L}}(\sigma_0)L}{\gamma_T(\sigma_0)}. \tag{3.13}$$

In order to explore whether this extremum is a minimum, note first from (3.3) that

$$\frac{\partial \bar{\sigma}(\delta, s)}{\partial s} = -M(\bar{\sigma}(\delta, s)) \gamma_T(\bar{\sigma}(\delta, s)), \tag{3.14}$$

where $\gamma_T(\sigma)$ is the transformation strain, and $M(\sigma, s)$ is the stiffness of the stress–elongation curve at fixed s:

$$M(\sigma, s) = \left. \frac{\partial \bar{\sigma}(\delta, s)}{\partial \delta} \right|_{\delta = \hat{\delta}(\sigma, s)} = 1/\left(\Gamma_{\mathsf{H}}'(\sigma)s + \Gamma_{\mathsf{L}}'(\sigma)(L - s) \right) > 0. \tag{3.15}$$

It now follows from (3.6), (3.14), and (3.15) that

$$\frac{\partial^2 \bar{E}(\delta, s)}{\partial s^2} = \left. \left(M(\sigma, s) \gamma_T^2(\sigma) \right) \right|_{\sigma = \bar{\sigma}(\delta, s)} > 0. \tag{3.16}$$

From (3.16) one can conclude that this extremum is in fact a minimum. We have therefore established the following result:

Among all mixed-phase equilibria of the form (3.1) and (3.3) with δ given, the one for which the stress in the bar coincides with the Maxwell stress minimizes the total energy for the given δ.

We might say that, among all the mixed-phase equilibria (3.3) with a given δ, the "Maxwell phase mixture" just described is "most stable" in an energy sense.

We next show that the energy-minimizing property of the Maxwell equilibrium phase mixture persists when the class of competing deformations is much larger than that encompassing only the mixed-phase strain distributions (3.1). To do this, we study the behavior of the total stored energy

$$E\{\tilde{u}\} = \int_0^L W(\tilde{u}'(x)) \, dx \tag{3.17}$$

as $\tilde{u}$, viewed as a candidate displacement field for our displacement boundary value problem, is varied over a suitable class of functions; recall that W is the strain energy per unit reference volume for the material of the bar. The class of admissible functions consists of those $\tilde{u}$'s that are continuous with piecewise continuous first and second derivatives on $[0, L]$ and satisfy the displacement boundary conditions:

$$\tilde{u}(0) = 0, \qquad \tilde{u}(L) = \delta. \tag{3.18}$$

We seek a $\tilde{u}$ that minimizes $E\{\tilde{u}\}$ over the class of admissible candidates; such a $\tilde{u}$ would be stable in a minimum-energy sense, not merely with respect to the mixed-phase competitors of (3.1), but with respect to all candidates in the enlarged class.

Suppose the minimum of $E\{\tilde{u}\}$ is attained at a function $\tilde{u} = u$, continuous on $[0, L]$, whose gradient $u'(x)$ has a discontinuity at $x = s$, but, together with $u''(x)$, is continuous elsewhere in $[0, L]$. Let $\gamma(x) = u'(x)$ and $\sigma(x) = \hat{\sigma}(\gamma(x))$, respectively, be the strain and stress formally associated with this minimizer, and we write $\gamma^\pm = \gamma(s\pm), \sigma^\pm = \sigma(s\pm)$ for the respective strains and stresses on either side of the jump at $x = s$.

To derive conditions on this u that are necessary if it is to be a minimizer, we study $E\{\tilde{u}\}$ for $\tilde{u}$'s that are close to u. Let $\tilde{u}(\cdot, \epsilon)$ be a one-parameter family of admissible functions on $[0, L]$ for which $\tilde{u}(x, 0) = u(x)$. Assume that, for some $\epsilon_0 > 0$, $\tilde{u}$ is continuous in (x, ϵ) on $[0, L] \times (-\epsilon_0, \epsilon_0)$, with first and second derivatives that are continuous there *except* across $x = \bar{s}(\epsilon)$, where $\bar{s}(\epsilon)$ is a smooth function of ϵ for which $\bar{s}(0) = s$. Thus in particular, $\tilde{u}$ is close to u if $|\epsilon|$ is small.

Because $\tilde{u}$ is continuous across $x = \bar{s}(\epsilon)$, the limiting value of $\tilde{u}$ on this curve is $\tilde{u}(\bar{s}(\epsilon), \epsilon)$. Differentiating this with respect to ϵ shows that the jumps in the first derivatives $\tilde{u}_x$ and $\tilde{u}_\epsilon$ at this discontinuity must satisfy

$$[\![\tilde{u}_\epsilon]\!] = -\bar{s}_\epsilon [\![\tilde{u}_x]\!]. \tag{3.19}$$

Note that (3.19) is the precise analog of the kinematic jump condition (2.16) with the parameter ϵ here playing the role played by the time t in that setting.

Next, let

$$\mathcal{E}(\epsilon) = \int_0^{\bar{s}(\epsilon)} W(\tilde{u}_x(x, \epsilon)) \, dx + \int_{\bar{s}(\epsilon)}^L W(\tilde{u}_x(x, \epsilon)) \, dx \tag{3.20}$$

be the one-parameter family of total energies associated with $\tilde{u}(x, \epsilon)$. Since by assumption, $E\{\tilde{u}\}$ has a minimum at $\tilde{u} = u$, $\mathcal{E}(\epsilon)$ has a minimum at $\epsilon = 0$, and therefore $\mathcal{E}'(0) = 0$. By making use of integration by parts, the restriction (3.19), and the fact that $\tilde{u}(x, \epsilon)$ satisfies the boundary conditions imposed on $u(x)$ for all

ϵ, one can establish the following formula for $\mathcal{E}'(0)$:

$$\mathcal{E}'(0) = -Q - R - S, \tag{3.21}$$

where

$$Q = \int_0^s \sigma'(x)\eta(x)\,dx + \int_s^L \sigma'(x)\eta(x)\,dx, \tag{3.22}$$

$$R = (\sigma^+ - \sigma^-)\frac{\eta(s+) + \eta(s-)}{2}, \tag{3.23}$$

$$S = \left\{ W(\gamma^+) - W(\gamma^-) - \frac{\sigma^+ + \sigma^-}{2}(\gamma^+ - \gamma^-) \right\} \bar{s}_\epsilon(0), \tag{3.24}$$

and we have set

$$\eta(x) = \tilde{u}_\epsilon(x, 0). \tag{3.25}$$

In the expressions for Q, R, and S, the number $\bar{s}_\epsilon(0)$ is arbitrary, as is the function $\eta(x)$, except that necessarily $\eta(0) = \eta(L) = 0$. Since η may be taken to vanish outside an arbitrary subinterval of $(0, s)$ and $\bar{s}_\epsilon(0)$ may be chosen to be zero, the requirement $\mathcal{E}'(0) = 0$ implies that necessarily $\sigma(x)$ is constant on $(0, s)$. Similarly, one shows that $\sigma(x)$ is constant on (s, L). It then follows from (3.22) that $Q = 0$. Continuing to choose $\bar{s}_\epsilon(0) = 0$, we conclude that $R = 0$, and it follows (since $\eta(s\pm)$ may be chosen arbitrarily) that $\sigma^+ = \sigma^-$, so that $\sigma(x)$ is indeed constant on $(0, L)$; thus $Q = R = 0$, and so from (3.21) and the vanishing of $\mathcal{E}'(0)$, one has $S = 0$ as well. Finally, the arbitrariness of $\bar{s}_\epsilon(0)$ allows us to conclude that the contents inside the braces on the right in (3.24) must vanish, which is to say that *the driving force acting on the phase boundary is zero*. This conclusion represents an instance of the so-called *Weierstrass–Erdmann corner condition* from the calculus of variations, which is discussed, for example, in the book by Gelfand and Fomin [8]; for a discussion in the setting of three-dimensional nonlinear elasticity, see [1].

We have thus proved the following result:

If u is a minimizer of the total energy that satisfies the given displacement boundary conditions and has a single phase boundary, then the following must hold:

(i) *The associated stress must be uniform in the bar, so that mechanical equilibrium prevails;*

(ii) *The associated static driving force acting on the phase boundary must vanish, so that phase equilibrium prevails.*

The fact that the static driving force vanishes implies that the stress in the bar is the Maxwell stress σ_0. Reference to Figure 3.1 shows that a necessary condition on the prescribed elongation δ for this to hold is

$$\Gamma_L(\sigma_0)L \leq \delta \leq \Gamma_H(\sigma_0)L \quad \text{(mixed-phase minimizer)}. \tag{3.26}$$

If it is further assumed that, in the minimizer with a single phase boundary, the high-strain phase is on the left, we immediately recover the value of s in (3.13). If δ is in the range (3.26), then s as given by (3.13) indeed lies in the interval $[0, L]$. Thus if there is a displacement field with the properties assumed above that answers to the given δ and minimizes the total energy, the location s of its phase boundary is fully determined. Moreover, such a minimizing displacement, if it exists, must be given by

$$u(x) = \begin{cases} \Gamma_H(\sigma_0)x, & 0 \leq x \leq s, \\ \Gamma_L(\sigma_0)(x - L) + \delta, & s \leq x \leq L. \end{cases} \tag{3.27}$$

Continuity of u at $x = s$ is assured by (3.13).

If, on the other hand, our single-phase-boundary minimizer were to have the high-strain phase on the *right*, formulas different from (3.13) and (3.27) would result. One would find, however, that the two minimizers possess the same *volume fractions* of each phase, so that the values of the energy at the two minimizers are the same.

Indeed, one might consider energy minimizers with more than one phase boundary. One would find that the volume fractions of the two phases in such a minimizer are the same as those for single-phase-boundary minimizers, so that the stored energies would not differ from that for one phase boundary. This need *not* be the case if surface energy at phase boundaries were taken into account.

Each member of the one-parameter family of two-phase solutions (3.2) and (3.3) to the displacement boundary value problem is consistent with the requirements of mechanical equilibrium. But among such solutions, there is only one that minimizes the total energy for the given δ, and this minimizer places the bar in phase equilibrium as well as in mechanical equilibrium. Thus the requirement of energy minimization is stronger than the requirement of mechanical equilibrium alone. This is not the case for a bar made of a single-phase elastic material.

If an energy-minimizing, mixed-phase solution exists, the prescribed elongation must lie in the range (3.26). It is clear from Figure 3.1 that, over a portion of this range of δ, single-phase solutions of the boundary value problem exist as well. Indeed, the displacement field

$$u(x) = \delta x/L, \tag{3.28}$$

satisfies the necessary conditions for a local energy minimizer that leaves the bar in mechanical equilibrium in either the low-strain phase or the high-strain phase if either

$$-L < \delta \leq \gamma_M L \quad \text{(low-strain phase minimizer)}, \tag{3.29}$$

or

$$\gamma_m L \leq \delta < \infty \qquad \text{(high-strain phase minimizer)}, \qquad (3.30)$$

respectively.

Given a δ for which *both* the two-phase *and* a single-phase solution exist, one may compare the total stored energies to determine an energetically preferred solution. Let $E_L(\delta)$ $E_H(\delta)$, and $E_{HL}(\delta)$ be the total energies in the low-strain, high-strain, and mixed-phase solutions, respectively. Direct calculation gives

$$E_L(\delta) = LW(\delta/L) = LW(\Gamma_L(\sigma_0)) + L\int_{\Gamma_L(\sigma_0)}^{\delta/L} \hat{\sigma}(\gamma)\,d\gamma, \qquad -1 \leq \delta/L \leq \gamma_M,$$
$$(3.31)$$

$$E_H(\delta) = LW(\delta/L) = LW(\Gamma_H(\sigma_0)) - L\int_{\delta/L}^{\Gamma_H(\sigma_0)} \hat{\sigma}(\gamma)\,d\gamma, \qquad \gamma_m \leq \delta/L < \infty,$$
$$(3.32)$$

$$E_{HL}(\delta) = LW(\Gamma_L(\sigma_0)) + \sigma_0(\delta - \Gamma_L(\sigma_0))L, \qquad \Gamma_L(\sigma_0) \leq \delta/L \leq \Gamma_H(\sigma_0).$$
$$(3.33)$$

In obtaining (3.33), we have made use of the fact that the static driving force vanishes at the Maxwell stress, and we have used (3.13) to eliminate s in favor of δ. From (3.31) and (3.33), one finds that

$$E_{HL}(\delta) - E_L(\delta) = L\left\{\sigma_0(\delta/L - \Gamma_L(\sigma_0)) - \int_{\Gamma_L(\sigma_0)}^{\delta/L} \hat{\sigma}(\gamma)\,d\gamma\right\} \qquad (3.34)$$

for $\Gamma_L(\sigma_0) \leq \delta/L \leq \gamma_M$, where the pure low-strain phase solution and the mixed-phase solution both exist. Similarly,

$$E_{HL}(\delta) - E_H(\delta) = -L\left\{\sigma_0(\Gamma_H(\sigma_0) - \delta/L) - \int_{\delta/L}^{\Gamma_H(\sigma_0)} \hat{\sigma}(\gamma)\,d\gamma\right\} \qquad (3.35)$$

for the appropriate range of δ. By comparing the areas represented by the two terms in the braces in (3.34) and in (3.35), one finds that the right sides of these two equations are both negative. Thus one has

$$E_{HL}(\delta) \leq E_L(\delta), \qquad E_{HL}(\delta) \leq E_H(\delta) \qquad (3.36)$$

in the respective ranges of δ where the comparisons are meaningful, so that the mixed-phase solution is energetically preferred over *either* single-phase solution whenever the latter is a competitor.

In summary, the energy-minimizing configuration of the bar is as follows: for $\delta/L \leq \Gamma_L(\sigma_0)$, it involves the low-strain phase only, for $\delta/L \geq \Gamma_H(\sigma_0)$ it involves only the high-strain phase, and in the intermediate range $\Gamma_L(\sigma_0) \leq \delta/L \leq \Gamma_H(\sigma_0)$ it involves an equilibrium mixture of phases at the Maxwell stress.[1]

---

[1] The bold curve in Figure 3.3, which we encounter in the next section, can be associated with these configurations.

3.4 Quasistatic processes

A quasistatic process is modeled here as a one-parameter family of equilibrium states, the parameter being time t. Specifically, we suppose that the stress and the elongation in the description of states of mechanical equilibrium are formally replaced by functions of time: $\sigma = \sigma(t)$, $\delta = \delta(t)$. In the case of a two-phase equilibrium state, the location of the phase boundary, the strains on either side of it, and the associated driving force all become functions of time as well: $s = s(t)$, $\gamma^{\pm} = \gamma^{\pm}(t)$, and $f = f(t)$.

We now introduce a natural macroscopic notion of *dissipation rate* $D(t)$ for a bar undergoing a quasistatic process:

$$D(t) = \sigma(t)\dot{\delta}(t) \ - \ \frac{d}{dt} \int_0^L W(\gamma(x, t))\, dx. \qquad (3.37)$$

Note that the kinetic energy of the bar, though included in the definition of dissipation rate in the dynamic setting of Chapter 2, is neglected here. By means of a direct calculation that makes use of the facts that (i) the stress is uniform in the bar at each instant in a quasistatic process and (ii) the displacement $u(x, t)$ satisfies $u(0, t) = 0$ and $u(L, t) = \delta(t)$, one shows that

$$D(t) = f(t)\dot{s}(t), \qquad (3.38)$$

where $f(t)$ is the driving force of (3.4)–(3.6):

$$f(t) = W\left(\gamma^+(t)\right) - W\left(\gamma^-(t)\right) - \sigma(t)\left(\gamma^+(t) - \gamma^-(t)\right) = \mathsf{f}(\sigma(t)). \qquad (3.39)$$

Since $f(t)$ vanishes if $\gamma^+(t) = \gamma^-(t)$, (3.38) also applies when the bar is in a single-phase state, asserting that $D(t) = 0$ in such a case.

The dissipation associated with quasistatically moving surfaces of strain discontinuity in three-dimensional, nonlinearly elastic solids was studied in [11].

As in the discussion of dynamics in Chapter 2, it is postulated here that the dissipation rate cannot be negative:

$$D(t) = f(t)\dot{s}(t) \geq 0. \qquad (3.40)$$

The quasistatic process $\delta = \delta(t)$, $\sigma = \sigma(t)$ describes a path in the flag of equilibrium states. As shown in Figure 3.2, the dissipation inequality (3.40) restricts the directions that may be taken by this path at each point in the flag. At a point $(\delta(t), \sigma(t))$ lying above the "Maxwell line" $\sigma = \sigma_0$, the driving force $f(t)$ is positive by (3.6); thus by (3.40), $\dot{s}(t) \geq 0$ at such a point, so $s(t)$ cannot decrease in time, and the low-strain phase cannot grow. Below the Maxwell line, $s(t)$ must be nonincreasing, so the high-strain phase cannot grow. Motion in either direction along a curve $s = $ constant is permitted by (3.40), since such a motion is dissipation free. For the same reason, motion in either direction along the Maxwell line $\sigma = \sigma_0$ is also permitted. The clusters of arrows in Figure 3.2 show the admissible directions of a path at various points in the flag of equilibrium states.

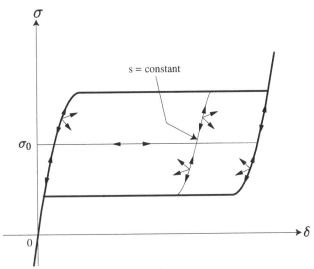

Figure 3.2. Arrows showing directions of paths of quasistatic processes that are permitted by the dissipation inequality (3.40); σ_0 is the Maxwell stress.

From (3.3), the elongation, stress, and phase boundary location during a quasistatic process are related through

$$\delta(t) = \Gamma_H(\sigma(t))s(t) + \Gamma_L(\sigma(t))(L - s(t)), \tag{3.41}$$

where $\gamma = \Gamma_H(\sigma)$ and $\gamma = \Gamma_L(\sigma)$ describe the respective high- and low-strain phases of the stress response function. In a displacement-controlled process, the elongation history $\delta(t)$ is given; observe that one cannot determine the corresponding stress history $\sigma(t)$ from (3.41), since the history of the phase boundary location $s(t)$ is not known.

One way to render the quasistatic process fully determinate would be to assume that the given elongation varies so slowly that the only states visited are those that are stable in the minimum-energy sense, so that mechanical equilibrium *and phase equilibrium* both prevail at all times. According to the discussion in the preceding section, this would impose the additional restriction $f(t) = 0$ on the process at each instant, which in turn would determine $s(t)$ according to the energy-minimizing formula (3.13), leading to the determination of $\sigma(t)$. Let us call this special quasistatic process the *Maxwell process*. It is clear from the comparison of energies of single-phase and mixed-phase solutions carried out at the end of the preceding section that the ultimate stress–elongation relation for the Maxwell process would be that corresponding to the bold curve in Figure 3.3. As the elongation is increased from zero, the stress increases along the low-strain phase branch of the stress–strain curve until the Maxwell stress is reached, at which time continued increase in δ leaves the stress at the value $\sigma = \sigma_0$ until the high-strain phase is reached. As δ is increased still further, the stress rises along the high-strain phase branch of the stress–strain

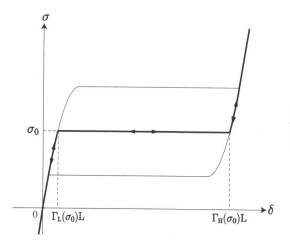

Figure 3.3. The stress–elongation curve for the dissipation-free Maxwell process is shown bold.

curve. If the bar is subsequently shortened, the unloading path is simply the reverse of the path followed on loading.

The Maxwell model just described is completely free of dissipation, since driving force vanishes at all times. Such a conservative theory is far too restrictive to describe observed dissipative phenomena such as hysteresis in a cycle of loading of the kind just sketched.

3.5 Nucleation and kinetics

To construct a model capable of describing dissipative response of two-phase non-linearly elastic materials, we first identify three questions concerning the phase transition itself that will help us proceed. First, at which particle or particles in the bar does the transition first occur? This is the issue of the *nucleation site*. Second, once the nucleation site is known, under what conditions will the transition in fact initiate there? This is the issue of the *nucleation criterion*. Finally, once the transition has nucleated at the specified site, how does it evolve? This is a matter of specifying the *kinetics* controlling the growth of the "product phase" at the expense of the "parent phase," or equivalently, the agent governing the speed of propagation of the phase boundary.

We adopt here the simplest answers to these questions. First, we address the issue of nucleation sites. Since our ideal bar is composed of a material that is homogeneous in the stress-free reference configuration, all particles away from the ends of the bar are indistinguishable. There is no reason to prefer one such particle over another in the absence of stress raisers such as voids or inclusions. The two particles at the ends, however, do differ from interior particles, and so we shall specify that nucleation always occurs at an end particle. We arbitrarily choose the particle at the end $x = 0$ of the bar as the nucleation site for the low-strain $\rightarrow$ high-strain phase transition, while the reverse transition is taken to nucleate at the particle $x = L$.

Second, when the entire bar is in one phase, a *necessary* condition for nucleation of a new phase is that the second phase be associated with a lower energy well than the existing phase. With reference to the potential energy function sketched in Figure 2.3, the low-strain phase is associated with the energy well on the left, while the high-strain phase is associated with the energy well on the right. The high-strain phase has less energy than the low-strain phase whenever $\sigma > \sigma_0$. However, this does not mean that the low-strain phase will necessarily transform into the high-strain phase when $\sigma > \sigma_0$, because there is an energy barrier between these two energy wells. The smaller the energy barrier, the more likely is the nucleation of the transformation. We shall take as the nucleation condition the criterion that a transformation will nucleate when the height of the energy barrier is at least as small as a certain critical value; this appears to be consistent with what is done in materials science, for example see Chapter 10 of Christian [6]. Since the height of the barrier can be expressed as a function of stress, which in turn can be expressed as a function of the driving force, we can equivalently state the nucleation criterion in terms of a critical value of driving force.

Therefore we postulate that nucleation of the low-strain $\rightarrow$ high-strain phase transition takes place when the static driving force f that would act on the incipient phase boundary is at least as great as a certain critical value f_n^{LH}. In view of (3.6), driving force increases monotonically with stress, so an equivalent version of the nucleation criterion in quasistatics asserts that nucleation of the forward transition occurs at the particle at $x = 0$ if the stress in the bar satisfies $\sigma \geq \sigma_n^{LH}$, where f_n^{LH} and σ_n^{LH} are related through (3.6). Similarly, the reverse transition at $x = L$ is associated with nucleation levels f_n^{HL} and σ_n^{HL} of driving force and stress, respectively. By (3.7), necessarily $0 \leq f_n^{LH} \leq f_M$ and $f_m \leq f_n^{HL} \leq 0$, and therefore the nucleation stresses must satisfy

$$\sigma_0 \leq \sigma_n^{LH} \leq \sigma_M, \qquad \sigma_m \leq \sigma_n^{HL} \leq \sigma_0. \qquad (3.42)$$

Often our discussion will be concerned only with the low-strain $\rightarrow$ high-strain, or forward, phase transition; when this is the case, we shall drop the superscript LH from the symbols for the nucleation parameters, writing instead simply f_n and σ_n.

We now turn to the kinetics of the transformation. Upon nucleation of the low-strain $\rightarrow$ high-strain transition, a phase boundary emerges from the end $x = 0$ of the bar and moves to the right; it has the high-strain phase on its left. We now assume that its referential speed $\dot{s}(t)$, which represents *the rate at which the material transforms from one phase to the other as it crosses the moving phase boundary, is determined by the local states on the two sides of the phase boundary* characterized by the pair γ^+, γ^-, that is, we assume that there is a "kinetic relation" of the form

$$\dot{s} = \tilde{v}(\gamma^+, \gamma^-). \qquad (3.43)$$

One can show that the equilibrium requirement $\hat{\sigma}(\gamma^+) = \hat{\sigma}(\gamma^-)$ and the definition (3.4) of driving force determine γ^+ and γ^- in terms of f. Thus the kinetic relation

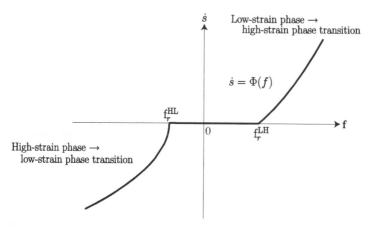

Figure 3.4. Schematic plot of a kinetic response function $\Phi(f)$. If the nucleation levels of driving force f_n^{LH} and f_n^{HL} were displayed on this figure, they would lie outside the interval (f_r^{HL}, f_r^{LH}) in keeping with the inequalities (3.47).

can be expressed in the form

$$\dot{s}(t) = \Phi(f(t)), \tag{3.44}$$

where Φ is a function that depends on the material. Except where expressly stated to the contrary, we shall assume throughout that $\Phi(f)$ increases monotonically with f, so that the kinetic relation (3.44) may be inverted:

$$f(t) = \phi(\dot{s}(t)), \tag{3.45}$$

where ϕ is the inverse of Φ. It is assumed that $\phi(\dot{s})$ is continuous for all $\dot{s} \neq 0$ and monotonically nondecreasing for all $\dot{s}$.

The dissipation inequality (3.40) restricts the kinetic response functions ϕ and Φ by requiring that

$$\phi(\dot{s})\dot{s} \geq 0, \qquad \Phi(f)f \geq 0. \tag{3.46}$$

A schematic plot of a generic kinetic response function Φ is shown in Figure 3.4. The portion of the graph for which $\Phi(f) < 0$ corresponds to $\dot{s} < 0$ and describes the kinetics of the high-strain $\rightarrow$ low-strain phase transition. The number $f_r^{LH} \geq 0$ shown in the figure represents the resistance that must be overcome to cause a phase boundary at rest to move in such a way that the high-strain phase grows; similarly the number $f_r^{HL} \leq 0$ corresponds to the growth of the low-strain phase. There is no loss of generality in assuming that the nucleation levels of driving force and the resistances to motion satisfy

$$f_n^{LH} \geq f_r^{LH} \geq 0, \qquad f_n^{HL} \leq f_r^{HL} \leq 0. \tag{3.47}$$

In principle, the form of the kinetic response function $\phi(\dot{s})$ or $\Phi(f)$ might be derived from modeling at the microscopic level. One instance of this is the kinetic relation based on thermal activation considerations at the atomic level constructed in [3]. A different kinetic relation results if one uses an augmented continuum model that adds both viscous and strain-gradient contributions to the elastic part of the stress; by studying smooth traveling waves in such a solid in a suitable limit that removes these augmenting effects, one finds the particular kinetic relation obtained in [4]. The effect of inhomogeneities on the propagation of phase boundaries in one-dimensional tensile bars has been modeled by Bhattacharya [5] and by Rosakis and Knowles [18]; see Section 8.4.3 for further details. These models predict irregular phase boundary motion of the kind sometimes seen in experiments [13].

Note that the dissipation-free Maxwell model governed by the requirement $f(t) = 0$ is subsumed under the present theory of kinetics; it corresponds to the choices $\phi = 0$ in (3.45) and $f_n^{LH} = f_n^{HL} = f_r^{LH} = f_r^{HL} = 0$ in (3.47).

Since f may be expressed in terms of σ according to (3.6), one may recast the kinetic relation in the form

$$\dot{s} = \bar{\Phi}(\sigma). \tag{3.48}$$

By eliminating the internal variable s between (3.3) and (3.48), one finds the "master differential equation" relating the histories of stress $\sigma(t)$ and elongation $\delta(t)$ for any loading process. It may be written as

$$\left[p(\sigma) - \gamma_T'(\sigma)\delta/L \right] \dot{\sigma} + \gamma_T(\sigma)\dot{\delta}/L = \gamma_T^2 \bar{\Phi}(\sigma)/L, \tag{3.49}$$

where

$$p(\sigma) = \Gamma_L(\sigma)\Gamma_H'(\sigma) - \Gamma_L'(\sigma)\Gamma_H(\sigma), \tag{3.50}$$

and $\gamma_T(\sigma) = \Gamma_H(\sigma) - \Gamma_L(\sigma)$ is the transformation strain. In (3.49), usually one would either specify $\sigma(t)$ and seek $\delta(t)$, corresponding to a load-controlled process, or $\delta(t)$ would be given, with $\sigma(t)$ to be determined, in a displacement-controlled loading program. Since the latter is the more common case in experiments of interest here, it is the one we shall consider.

3.6 Constant-elongation rate processes

To illustrate the preceding discussion and calculate the response of the bar corresponding to a specific loading program, we now consider a process in which the bar is first extended from its stress-free, low-strain-phase state at a given constant-elongation rate $\dot{\delta} > 0$ (loading) and then shortened at the elongation rate $-\dot{\delta}$ (unloading).

In order to keep the details as simple as possible, we restrict attention to the special case of the *trilinear* two-phase material, in which all three branches of the stress–strain curve are straight lines, with the two rising branches given by

$$\hat{\sigma}(\gamma) = \begin{cases} \mu\gamma, & -1 < \gamma \leq \gamma_M, \\ \mu(\gamma - \gamma_T), & \gamma \geq \gamma_m, \end{cases} \tag{3.51}$$

where $\mu > 0$ is the elastic modulus, taken to be the same in both phases for simplicity, and the transformation strain $\gamma_T > 0$ is constant; see Figure 3.5.

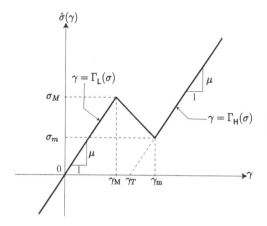

Figure 3.5. Trilinear material with equal moduli μ in both phases and constant transformation strain γ_T.

It follows that

$$\Gamma_L(\sigma) = \sigma/\mu, \qquad \Gamma_H(\sigma) = \gamma_T + \sigma/\mu. \tag{3.52}$$

The Maxwell stress is given by $\sigma_0 = (\sigma_M + \sigma_m)/2$, and the representation (3.6) for the static driving force becomes

$$f = (\sigma - \sigma_0)\gamma_T. \tag{3.53}$$

As the bar elongates at the given constant rate $\dot{\delta}$, the entire bar is initially in the low-strain phase and so the stress and elongation are related by

$$\delta = L\Gamma_L(\sigma) = L\sigma/\mu \quad \text{(loading: low-strain phase)}. \tag{3.54}$$

The stress rises up the left flagpole and eventually reaches the critical value σ_n^{LH} at which time the high-strain phase nucleates at $x = 0$. The elongation at this instant is $\delta_n^{LH} = L\Gamma_L(\sigma_n^{LH}) = L\sigma_n^{LH}/\mu$; see Figure 3.6. During the next stage of loading the bar is in a two-phase state. As the elongation continues to increase, the phase

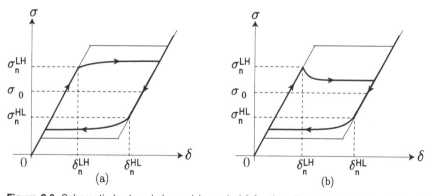

Figure 3.6. Schematic hysteresis loops (shown bold) for the trilinear material. The high-strain phase is nucleated during loading at $(\delta_n^{LH}, \sigma_n^{LH})$, and the low-strain phase is nucleated during unloading at $(\delta_n^{HL}, \sigma_n^{HL})$. The bar is fully transformed prior to unloading. Figures (a) and (b) correspond to, respectively, sufficiently fast and sufficiently slow loading rates, and they illustrate load-rise and load-drop at nucleation during loading.

boundary propagates rightwards according to the kinetic relation, transforming material from the low-strain to the high-strain phase. In describing the macroscopic response of the bar, we may regard σ as a function of δ, so that $\dot{\sigma} = \dot{\delta}\, d\sigma/d\delta$ since $\delta = \dot{\delta}\, t$. The stress–elongation relation during this two-phase portion of the response is then the solution of the initial-value problem

$$\left.\begin{aligned}
\frac{d\sigma}{d\delta} + (\mu\gamma_T/\dot{\delta}L)\bar{\Phi}(\sigma) &= \mu/L, \\[2mm]
\sigma = \sigma_n^{LH} \quad \text{at} \quad \delta &= \delta_n^{LH},
\end{aligned}\right\} \qquad \text{(two-phase loading)}, \qquad (3.55)$$

where $(3.55)_1$ follows from the master equation (3.49) specialized to the present case. Observe that the only nonlinearity in this initial-value problem (for the trilinear stress–strain relation) arises from the kinetic relation.

Before continuing to the next stage of loading, we examine the behavior of the stress immediately following nucleation of the high-strain phase, with particular attention to whether the stress rises or falls. By setting $\sigma = \sigma_n^{LH}$ and $\delta = \delta_n^{LH}$ in (3.55), the slope of the stress–elongation curve immediately after nucleation is seen to be

$$\left.\frac{d\sigma}{d\delta}\right|_{\delta=\delta_n^{LH}} = \frac{\mu}{L}\left(1 - \gamma_T\bar{\Phi}(\sigma_n^{LH})/\dot{\delta}\right). \qquad (3.56)$$

If the nucleation stress σ_n^{LH} and the resistance stress σ_r^{LH} for phase boundary motion coincide, (3.56) shows that the initial slope is positive since $\bar{\Phi}(\sigma_r^{LH}) = 0$. On the other hand, if $\sigma_n^{LH} > \sigma_r^{LH}$, one has

$$\left.\frac{d\sigma}{d\delta}\right|_{\delta=\delta_n^{LH}} \begin{cases} > 0 \quad \text{for} \quad \dot{\delta} > \gamma_T\bar{\Phi}(\sigma_n^{LH}) \quad \text{(``load rise'')}, \\[2mm] < 0 \quad \text{for} \quad \dot{\delta} < \gamma_T\bar{\Phi}(\sigma_n^{LH}) \quad \text{(``load drop'')}. \end{cases} \qquad (3.57)$$

Thus if $\sigma_n^{LH} > \sigma_r^{LH}$, the stress may decrease after nucleation if the elongation rate $\dot{\delta}$ is slow enough; if $\sigma_n^{LH} = \sigma_r^{LH}$, this cannot occur. Figures 3.6(a) and (b) depict two different elongation rates to illustrate load rise and load drop.

The differential equation in (3.55) holds until the point $(\delta(t), \sigma(t))$ reaches the right flagpole. At this instant the phase boundary has traversed the entire length of the specimen and the bar is now entirely in the high-strain phase. Continued loading causes the response to follow up the right flagpole and therefore is described by

$$\delta = \Gamma_H(\sigma)L = L\sigma/\mu + L\gamma_T \quad \text{(loading: high-strain phase)}. \qquad (3.58)$$

Neither the kinetic relation nor the nucleation condition is now relevant. If the loading is stopped and reversed at a subsequent time, the response first follows down the right flagpole, eventually reaching the critical value σ_n^{HL} at which time the low-strain phase nucleates at $x = L$. The elongation at this instant is $\delta_n^{HL} = \Gamma_H(\sigma_n^{HL})L = L\sigma_n^{LH}/\mu + L\gamma_T$; see Figure 3.6. During continued unloading, the low-strain phase grows at the expense of the high-strain phase as the phase boundary moves leftwards, the kinetic relation being in force again. By specializing the master equation (3.49), this unloading process can be described by the following initial-value problem:

$$
\left.\begin{aligned}
\frac{d\sigma}{d\delta} - (\mu\gamma_T/\dot\delta L)\bar\Phi(\sigma) &= \mu/L \\[6pt]
\sigma = \sigma_n^{HL} \quad \text{at} \quad \delta &= \delta_n^{HL}
\end{aligned}\right\} \qquad \text{(two-phase unloading).} \qquad (3.59)
$$

Eventually the point (δ, σ) returns to the left flagpole, the entire bar having returned to the low-strain phase.

Perhaps it is worth emphasizing that in the context of a uniform bar, the nucleation criterion is triggered only when the bar is in a single-phase state and is seeking to initiate a phase transition; once the bar is in a two-phase state, the nucleation criterion is no longer relevant, and it is the kinetic relation that now controls the evolution of the two-phase state. In the scenario described above, the nucleation criterion is triggered twice, once during loading and once during unloading; the kinetic relation is in force whenever a phase boundary is present in the bar.

To simplify the results even further, consider the particular kinetic relation

$$
\dot s = \Phi(f) = \begin{cases}
\mathsf{M}(f - f_r) & \text{if} \quad f > f_r, \\[4pt]
0 & \text{if} \quad -f_r \le f \le f_r, \\[4pt]
\mathsf{M}(f + f_r) & \text{if} \quad f < -f_r.
\end{cases} \qquad (3.60)
$$

Here the constant $\mathsf{M} > 0$ is the phase boundary *mobility*; its units are those of velocity per unit stress. Moreover, we have taken $f_r^{LH} = -f_r^{HL} = f_r$ so that by (3.53), the corresponding values of stress are $\sigma_r^{LH} = \sigma_0 + f_r/\gamma_T$ and $\sigma_r^{HL} = \sigma_0 - f_r/\gamma_T$. In terms of stress, the kinetic response function corresponding to this $\Phi(f)$ is given, according to (3.53) and (3.60), by

$$
\bar\Phi(\sigma) = \begin{cases}
\mathsf{M}\gamma_T(\sigma - \sigma_r^{LH}) & \text{if} \quad \sigma > \sigma_r^{LH}, \\[4pt]
0 & \text{if} \quad \sigma_r^{HL} \le \sigma \le \sigma_r^{LH}, \\[4pt]
\mathsf{M}\gamma_T(\sigma - \sigma_r^{HL}) & \text{if} \quad \sigma < \sigma_r^{HL}.
\end{cases} \qquad (3.61)
$$

For the trilinear material and the kinetic relation (3.60), the initial-value problem (3.55) corresponding to two-phase loading has the following solution:

$$
\sigma = \sigma_r^{LH} + \mu v + \left(\sigma_n^{LH} - \sigma_r^{LH} - \mu v\right) \exp\left(-\frac{\delta - \delta_n^{LH}}{vL}\right) \qquad \text{(two-phase loading),}
$$
$$(3.62)$$

and the initial-value problem (3.59) for two-phase unloading has the solution

$$
\sigma = \sigma_r^{HL} - \mu v + \left(\sigma_n^{HL} - \sigma_r^{HL} + \mu v\right) \exp\left(\frac{\delta - \delta_n^{HL}}{vL}\right) \qquad \text{(two-phase unloading).}
$$
$$(3.63)$$

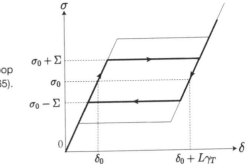

Figure 3.7. Rate-independent hysteresis loop (shown bold) for the frictional kinetic relation (3.65). The bar is fully transformed prior to unloading.

Here the dimensionless elongation rate ν is defined by

$$\nu = \frac{\dot{\delta}}{\mu M \gamma_T^2}. \tag{3.64}$$

Parenthetically, note that in this special case, load drop at nucleation occurs during loading if $\dot{\delta} < M\gamma_T^2(\sigma_n^{LH} - \sigma_r^{LH})$.

A special case of interest results if one assumes that, in the kinetic relation (3.60), the resistance f_r to phase boundary motion coincides with the nucleation value f_n of driving force for both forward and reverse transitions and then takes the limit as the mobility $M \to \infty$. The kinetic relation that arises from applying this process formally to (3.60) is the "frictional" one described by

$$f = \phi(\dot{s}) = \begin{cases} f_n & \text{if } \dot{s} > 0, \\ -f_n & \text{if } \dot{s} < 0. \end{cases} \tag{3.65}$$

Upon taking the limit, the two-phase portions (3.62) and (3.63) of the stress–elongation response become

$$\sigma = \begin{cases} \sigma_0 + \Sigma & \text{if } \delta_0 + \Delta \leq \delta \leq \delta_0 + \Delta + \gamma_T L \quad \text{(two-phase loading)}, \\ \\ \sigma_0 - \Sigma & \text{if } \delta_0 - \Delta \leq \delta \leq \delta_0 - \Delta + \gamma_T L \quad \text{(two-phase unloading)}, \end{cases} \tag{3.66}$$

where $\Sigma = \sigma_n^{LH} - \sigma_0$, $\Delta = L\Sigma/\mu$, and $\delta_0 = L\sigma_0/\mu$. The parallelogram in the δ, σ-plane formed by the two horizontal lines described by (3.66), together with the parts of the trilinear flagpoles cut off by these lines, comprises the hysteresis loop determined by the frictional kinetic relation (3.65); see Figure 3.7. This hysteresis loop is independent of the elongation rate.

Finally we consider a loading program that is identical to the preceding one *except that* now, loading is stopped and reversed *before* the phase boundary reaches $x = L$. In contrast to the previous program, here, unloading commences while the bar is still in a two-phase state and therefore the nucleation criterion plays no role during unloading. This is different to the first loading program where unloading

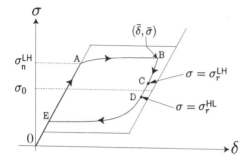

Figure 3.8. Hysteresis loop (shown bold) for the trilinear material. Unloading commences before the entire bar has transformed to the high-strain phase.

commenced from a single-phase state and therefore involved the creation of a new interface according to the nucleation criterion.

The response of the bar from the initial instant until the time $\bar{t}$ when the loading is reversed remains the same as in the previous program. The elongation $\bar{\delta}$ at $t = \bar{t}$, that is, the maximum elongation, is prescribed. The corresponding stress $\bar{\sigma} = \sigma(\bar{\delta})$ must be such that the point $(\bar{\delta}, \sigma(\bar{\delta}))$ lies in the *interior* of the flag and, of course, above the Maxwell line.

The unloading process begins from $t = \bar{t}$, $\delta = \bar{\delta}$, $\sigma = \bar{\sigma}$ (point B in Figure 3.8) and, according to the master equation (3.49), is described by the following initial-value problem:

$$
\left.
\begin{aligned}
\frac{d\sigma}{d\delta} - (\mu\gamma_T/\dot{\delta}L)\bar{\Phi}(\sigma) &= \mu/L \\[2mm]
\sigma = \bar{\sigma} \quad \text{at} \quad \delta = \bar{\delta}
\end{aligned}
\right\}
\quad \text{(unloading).} \qquad (3.67)
$$

As δ decreases, so does σ. Recall that the kinetic response function $\bar{\Phi}(\sigma)$ involves three distinct ranges $\sigma > \sigma_r^{LH}$, $\sigma_r^{HL} \leq \sigma \leq \sigma_r^{LH}$, and $\sigma < \sigma_r^{HL}$, corresponding to the three branches of $\Phi(f)$ in Figure 3.4, which by the kinetic relation correspond to $\dot{s} > 0$, $\dot{s} = 0$, and $\dot{s} < 0$ respectively. As the stress decreases from $\bar{\sigma}$ to 0 during unloading, it passes through each of these ranges in sequence. First, as the stress decreases from $\bar{\sigma}$ to σ_r^{LH} we have $\dot{s} > 0$, and so the phase boundary continues to move to the right even though the elongation and stress are both decreasing; the corresponding unloading path is the curve BC in Figure 3.8. (This feature is not peculiar to the particular kinetic relation (3.60) since, for any kinetic law, the dissipation inequality will not permit the phase boundary to reverse its direction of travel until the Maxwell line is crossed.) In the next stage, as the stress decreases from σ_r^{LH} to σ_r^{HL}, the phase boundary remains stationary ($\dot{s} = 0$). The corresponding portion of the graph of $\sigma = \sigma(\delta)$ therefore lies on a line s=constant in the flag of equilibrium states – the straight line segment CD in Figure 3.8. After the stress passes through σ_r^{HL}, the phase boundary moves to the left, and the point (δ, σ) eventually returns to the left flagpole to complete the hysteresis loop.

When specialized to the kinetic relation (3.60), the initial-value problem (3.67) for unloading is readily solved. The response involves three stages as described above. Immediately following the reversal of the sign of $\dot\delta$, one has

$$\sigma = \sigma_r^{\mathrm{LH}} - \mu\nu + \left(\mu\nu + \bar\sigma - \sigma_r^{\mathrm{LH}}\right) \exp\left(\frac{\delta - \bar\delta}{\nu L}\right),$$

$$\delta_r^{\mathrm{LH}} \le \delta \le \bar\delta \quad \text{(unloading, stage 1)};$$

(3.68)

here δ_r^{LH} is the value of elongation when the stress has diminished to the value σ_r^{LH}; it is found from

$$\exp\left(\frac{\delta_r^{\mathrm{LH}} - \bar\delta}{\nu L}\right) = \frac{\mu\nu}{\mu\nu + \bar\sigma - \sigma_r^{\mathrm{LH}}}.$$

(3.69)

In the second stage of unloading the path in the δ, σ-plane is the $s = \text{constant}$ straight line segment

$$\sigma = \sigma_r^{\mathrm{LH}} + \frac{\mu}{L}(\delta - \delta_r^{\mathrm{LH}}), \qquad \delta_r^{\mathrm{HL}} \le \delta \le \delta_r^{\mathrm{LH}} \quad \text{(unloading, stage 2)}, \quad (3.70)$$

where

$$\delta_r^{\mathrm{HL}} = \delta_r^{\mathrm{LH}} - \frac{L}{\mu}(\sigma_r^{\mathrm{LH}} - \sigma_r^{\mathrm{HL}}).$$

(3.71)

In the third stage of unloading, the path moves from the point $(\delta_r^{\mathrm{HL}}, \sigma_r^{\mathrm{HL}})$ to the low-strain flagpole; one finds that

$$\sigma = \sigma_r^{\mathrm{HL}} + \mu\nu\left(\exp\left(\frac{\delta - \delta_r^{\mathrm{HL}}}{\nu L}\right) - 1\right), \quad \delta_{\mathrm{final}} \le \delta \le \delta_r^{\mathrm{HL}} \quad \text{(unloading, stage 3)},$$

(3.72)

where δ_{final}, the value of the elongation when the flagpole is reached, is the root of

$$\delta_{\mathrm{final}} = \frac{\sigma_r^{\mathrm{HL}}}{\mu}L + \nu\left(\exp\frac{(\delta_{\mathrm{final}} - \delta_r^{\mathrm{HL}})}{\nu L} - 1\right).$$

(3.73)

The stress–elongation curves arising from (3.62) and (3.68)–(3.73) provide the hysteresis loop. Examples are shown in Figures 3.8 and 3.9.

Figure 3.9 refers to a trilinear material for which $\sigma_m < 0$, so that permanent deformation can occur on unloading.

3.7 Hysteresis

Our objective in this section is to study the influence of the kinetic relation on the hysteresis associated with constant-elongation rate cycles, assuming the material to be trilinear, but allowing the kinetic response function $\bar\Phi(\sigma)$ to be reasonably general. Under certain assumptions, we show that, as observed in some experiments, the hysteresis H as predicted by the model is an increasing function of the elongation rate $\dot\delta$ for the class of kinetic relations considered. We then adopt a particular kinetic relation and calculate the dependence of H on $\dot\delta$ numerically. Finally, for this particular kinetic relation, we show that, as $\dot\delta$ tends to zero, the limiting value

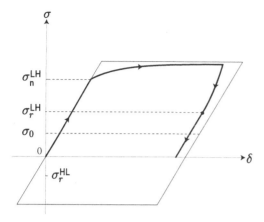

Figure 3.9. Hysteresis loop (shown bold) for a trilinear material (3.51) for which the stress at the local minimum is negative: $\sigma_m < 0$. Unloading results in permanent deformation at zero stress. The dimensionless elongation rate is $\nu = 1$.

of H is zero only under very restrictive circumstances, so that *static hysteresis* [9] normally occurs.

We make several inessential concessions to simplicity:

(i) The kinetic response function $\Phi(f)$ and its inverse $\phi(\dot{s})$ are assumed to be odd functions of their arguments;

(ii) For both forward and reverse phase transitions, the nucleation level of driving force coincides with the resistance to phase boundary motion;

(iii) Loading continues until the bar is fully transformed to the high-strain phase, at which time unloading begins and continues until the bar is fully transformed to the low-strain phase.

Under these assumptions, the total energy H dissipated during the cycle is twice the energy lost during loading alone, so that

$$H = 2 \int_0^\tau D(t)\, dt = 2 \int_0^\tau f(t)\dot{s}(t)\, dt = 2 \int_0^\tau f(t)\Phi(f(t))\, dt, \quad (3.74)$$

where $D(t)$ is the dissipation rate at time t, and τ is the time required to fully transform the bar to the high-strain phase during loading. Since this is the time required for the phase boundary to reach the far end of the bar, (3.44) yields an equation to determine τ:

$$L = \int_0^\tau \Phi(f(t))dt. \qquad (3.75)$$

For present purposes, it is convenient to use (3.53) to rewrite the initial-value problem (3.55) for loading as a problem for f, rather than σ:

$$\left. \begin{aligned} \dot{f} &= (\mu\gamma_T/L)\left(\dot{\delta} - \gamma_T\Phi(f)\right), \\ f &= f_n \quad \text{at } t = 0. \end{aligned} \right\} \qquad (3.76)$$

Let $\bar{f} = \bar{f}(\dot{\delta})$ be the unique zero of $\dot{\delta} - \gamma_T\Phi(f)$; then $f(t)$ increases with t if $f_n < \bar{f}$ (load rise), but decreases if $f_n > \bar{f}$ (load drop). In either case, $f(t)$ is monotonic in t, so that one may use the differential equation $(3.76)_1$ to rewrite (3.75) as follows:

$$\mu \gamma_T = \int_{f_n}^{f_\tau} \frac{\Phi(f)}{\dot{\delta} - \gamma_T \Phi(f)} df, \tag{3.77}$$

where f_τ is the value of driving force at the time τ of arrival of the phase boundary at $x = L$. Moreover, since $D(t) = f(t)\dot{s}(t) = f(t)\Phi(f(t))$, we may also rewrite (3.74) in the form

$$H = 2\frac{L}{\mu \gamma_T} \int_{f_n}^{f_\tau} \frac{f \Phi(f)}{\dot{\delta} - \gamma_T \Phi(f)} df. \tag{3.78}$$

Thus for a given elongation rate $\dot{\delta}$, one determines $f_\tau = f_\tau(\dot{\delta})$ from (3.77) and then uses this value of f_τ in (3.78) to find the total dissipation associated with the hysteresis loop.

From (3.77), one can show that

$$\frac{df_\tau(\dot{\delta})}{d\dot{\delta}} = \frac{\dot{\delta} - \gamma_T \Phi(f_\tau)}{\Phi(f_\tau)} \int_{f_n}^{f_\tau} \frac{\Phi(f)df}{(\dot{\delta} - \gamma_T \Phi(f))^2}. \tag{3.79}$$

Since $\dot{\delta} - \gamma_T \Phi(f_\tau)$ and $f_\tau - f_n$ have the same sign in both the load-drop and load-rise cases, it follows from (3.79) that $df_\tau(\dot{\delta})/d\dot{\delta} > 0$ for both cases, so that $f_\tau(\dot{\delta})$ is always an increasing function of $\dot{\delta}$. Moreover, (3.78) and (3.79) together lead to

$$\frac{dH}{d\dot{\delta}} = \frac{L}{\mu \gamma_T} \int_{f_n}^{f_\tau} \frac{(f_\tau - f)\Phi(f)df}{(\dot{\delta} - \gamma_T \Phi(f))^2} > 0 \tag{3.80}$$

for both the load-rise and load-drop cases. Thus according to the present model, the hysteresis H is always an *increasing* function of the elongation rate $\dot{\delta}$.

In the experiments on NiTi shape-memory wires carried out by Lin et al. [14], the hysteresis is found to be an increasing function of relative elongation rate over the two decades of $\dot{\delta}/L$ from 1% per minute to 100% per minute at each of three different temperatures. In practice, the thermal effects associated with phase transformation can become significant at higher elongation rates, and so the experimentally observed increase in hysteresis is presumably due to both thermal and kinetic effects. This will be examined further in Chapter 10.

In the case of the special kinetic relation (3.60), the integration in (3.77) can be carried out explicitly, furnishing the following equation for the determination of $f_\tau = f_\tau(\dot{\delta})$:

$$f_\tau - f_n + \frac{\dot{\delta}}{\gamma_T \mathsf{M}} \log \left| \frac{f_\tau - f_r - \dot{\delta}/(\gamma_T \mathsf{M})}{f_n - f_r - \dot{\delta}/(\gamma_T \mathsf{M})} \right| + \mu \gamma_T^2 = 0. \tag{3.81}$$

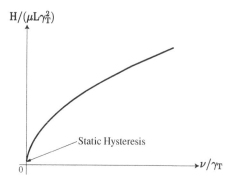

H/($\mu L \gamma_T^2$)

Static Hysteresis

0

ν/γ_T

Figure 3.10. Dimensionless total dissipation $H/(\mu L \gamma_T^2)$ of (3.85) as a function of dimensionless elongation rate ν/γ_T. It was assumed that $\gamma_r = \gamma_n$, and the values of $\gamma_M = 0.8$, $\gamma_m = 1.3$, $\gamma_T = 1$, $\gamma_n = 0.7$ were used in the calculation. The nonzero limiting value of the total dissipation as $\nu \to 0$ is shown in the figure as *static hysteresis*.

With the help of (3.81), one finds that the representation (3.78) for the total dissipation H specializes to

$$H/(2L) = f_r + \frac{\dot{\delta}}{\gamma_T \mathsf{M}} + \frac{1}{2\mu\gamma_T^2}\left[(f_n - f_r)^2 - (f_\tau(\dot{\delta}) - f_r)^2\right]. \quad (3.82)$$

Both (3.81) and (3.82) are valid for both the load-rise and load-drop cases.

To use (3.81) and (3.82) to compute the hysteresis H as a function of $\dot{\delta}$ for a given material, it is helpful to rewrite both equations in terms of dimensionless quantities in such a way as to make clear the number of independent parameters that must be specified. To this end, we use the dimensionless elongation rate ν introduced in (3.64), and we make use of the low-strain-phase strain γ_τ corresponding to the stress σ_τ in the bar at the instant $t = \tau$ when the phase boundary arrives at $x = L$. With the help of the relation (3.53) between stress and driving force, we write

$$\gamma_\tau = \sigma_\tau/\mu = \sigma_0/\mu + f_\tau/(\mu\gamma_T) = \gamma_0 + f_\tau/(\mu\gamma_T), \quad (3.83)$$

where $\gamma_0 = \sigma_0/\mu$ is the low-strain-phase Maxwell strain. Using (3.64) and (3.83) in (3.81) and (3.82) provides the equation to determine γ_τ and the corresponding expression for the hysteresis:

$$\gamma_\tau - \gamma_n + \nu \log\left|\frac{\gamma_\tau - \gamma_r - \nu}{\gamma_n - \gamma_r - \nu}\right| = -\gamma_T, \quad (3.84)$$

$$\frac{H}{\mu L} = (\gamma_\tau - \gamma_n)^2 - 2(\gamma_\tau - \gamma_n)(\gamma_\tau - \gamma_r) + 2\gamma_T(\nu + \gamma_r - \gamma_0). \quad (3.85)$$

For conveniently chosen but otherwise meaningless values of certain material parameters, Figure 3.10 plots the dimensionless total dissipation $H/(\mu L \gamma_T^2)$ as a function of a dimensionless elongation rate ν/γ_T in the case $\gamma_r = \gamma_n$.

From (3.82), one can show that, as the elongation rate tends to zero, the limiting value of total dissipation in a loading cycle is given by

$$H_{\text{static}} = \lim_{\nu \to 0} H = 2L\mu\left\{\gamma_T(\gamma_r - \gamma_0) + (1/2)(\gamma_n - \gamma_r)^2\right\}; \quad (3.86)$$

this limit is the counterpart in the present circumstances of what is called *static hysteresis* in [9]. It may be noted from (3.86) that, since $\gamma_T > 0$ and $\gamma_r \geq \gamma_0$, the

static hysteresis vanishes only under the very special conditions $\gamma_n = \gamma_r = \gamma_0$; this means that $\sigma_n = \sigma_r = \sigma_0$ and hence that $f_n = f_r = 0$. Thus static hysteresis is absent if and only if there is no resistance to phase boundary motion ($f_r = 0$) and nucleation takes place at the Maxwell stress σ_0 ($f_n = 0$). In this very special case, there is no interval on which $\dot{s} = 0$ in the kinetic relation (3.60); moreover, reference to (3.62) shows that, in the formal limit as loading rate tends to zero, the stress remains constant at the Maxwell stress as the phase boundary traverses the bar on loading. Under our assumption that nucleation of the reverse transition occurs when $f = -f_n$, the stress will also coincide with the Maxwell stress upon unloading, and hysteresis will indeed be absent under these special conditions.

For large elongation rates v, one can show that $H/(\mu L) = O(v^{1/2})$.

There are many experimental studies of the effects of cyclic loading on shape-memory materials; in particular, for nickel-titanium, a sample of such studies may be found in [10, 15, 16, 22]. Analytical models concerned with cyclic loading in such materials are given in [2, 12, 20, 21].

REFERENCES

[1] R. Abeyaratne, An admissibility condition for equilibrium shocks in finite elasticity. *Journal of Elasticity,* **13** (1983), pp. 175–84.

[2] R. Abeyaratne and S.-J. Kim, Cyclic effects in shape-memory alloys: a one-dimensional continuum model. *International Journal of Solids and Structures*, **34** (1997), pp. 3273–89.

[3] R. Abeyaratne and J.K. Knowles, A continuum model of a thermoelastic solid capable of undergoing phase transitions. *Journal of the Mechanics and Physics of Solids*, **41** (1993), pp. 541–71.

[4] R. Abeyaratne and J.K. Knowles, Implications of viscosity and strain-gradient effects for the kinetics of propagating phase boundaries in solids. *SIAM Journal on Applied Mathematics*, **51** (1991), pp. 1205–21.

[5] K. Bhattacharya, Phase boundary propagation in heterogeneous bodies. *Proceedings of the Royal Society of London* A, **455** (1999), pp. 757–66.

[6] J.W. Christian, *The Theory of Martensitic Transformations in Metals and Alloys*, Part 1. Pergamon, New York, 1975.

[7] J.L. Ericksen, Equilibrium of bars. *Journal of Elasticity*, **5** (1975), pp. 191–201.

[8] I.M. Gelfand and S.V. Fomin, *Calculus of Variations.* Prentice-Hall, New York, 1963.

[9] G.H. Goldsztein, F. Broner, and S.H. Strogatz, Dynamical hysteresis without static hysteresis: scaling laws and asymptotic expansions. *SIAM Journal on Applied Mathematics*, **57** (1997), pp. 1163–87.

[10] M. Kawaguchi, Y. Ohashi, and H. Tobushi, Cyclic characteristics of pseudoelasticity of Ti–Ni alloys. *JSME International Journal* A, **34** (1991), pp. 76–82.

[11] J.K. Knowles, On the dissipation associated with equilibrium shocks in finite elasticity. *Journal of Elasticity*, **9** (1979), pp. 131–58.

[12] J.K. Knowles, Hysteresis in the stress-cycling of bars undergoing solid–solid phase transitions. *Quarterly Journal of Mechanics and Applied Mathematics*, **55** (2002), pp. 69–91.

[13] R.V. Krishnan and L.C. Brown, Pseudo-elasticity and the strain–memory effect in an Ag-45 at. pct. Cd alloy. *Metallurgical Transactions* A, **4** (1973), pp. 423–9.

[14] P. Lin, H. Tobushi, K. Tanaka, T. Hattori, and A. Ikai, Influence of strain rate on deformation properties of TiNi shape memory alloy. *JSME International Journal* A, **39** (1996), pp. 117–23.

[15] K.N. Melton and O. Mercier, Fatigue of NiTi thermoelastic martensites. *Acta Metallurgica*, **27** (1979), pp. 137–44.

[16] S. Miyazaki, T. Imai, Y. Igo, and K. Otsuka, Effect of cyclic deformation on the pseudo-elasticity of Ti–Ni alloys. *Metallurgical Transactions* A, **17** (1986), pp. 115–20.

[17] N. Nakanishi, Lattice softening and the origin of SME, in *Shape-Memory Effects in Alloys*, edited by J. Perkins. Plenum Press, New York, 1975, pp. 147–76.

[18] P. Rosakis and J.K. Knowles, Continuum models for irregular phase boundary motion in solids. *European Journal of Mechanics* A, **18** (1999), pp. 1–16.

[19] J.A. Shaw and S. Kyriakides, Thermomechanical aspects of NiTi. *Journal of the Mechanics and Physics of Solids*, **43** (1995), pp. 1243–81.

[20] K. Tanaka, T. Hayashi, Y. Itoh, and H. Tobushi, Analysis of thermomechanical behavior of shape memory alloys. *Mechanics of Materials*, **13** (1992), pp. 207–15.

[21] K. Tanaka, F. Nishimura, T. Hayashi, H. Tobushi, and C. Lexcellent, Phenomenological analysis on subloops and cyclic behavior in shape memory alloys under mechanical and/or thermal loads. *Mechanics of Materials*, **19** (1995), pp. 281–92.

[22] H. Tobushi, P.-H. Lin, T. Hattori, and M. Makita, Cyclic deformation of TiNi shape memory alloy. *JSME International Journal* A, **38** (1995) pp. 59–67.

4 Impact-Induced Transitions in Two-Phase Elastic Materials

4.1 Introduction

We next turn to the dynamics of the two-phase nonlinearly elastic materials introduced in Chapter 2. As in the theory of mixed-phase equilibria and quasistatic processes for such materials set out in Chapter 3, the notion of driving force plays a central role in the analysis when inertial effects are taken into account. The indeterminacy exhibited in Chapter 3 by even the simplest static or quasistatic problems for two-phase materials manifests itself again in the present much richer dynamical context. Moreover, the continuum-mechanical interpretations of nucleation and kinetics again serve to restore the uniqueness of solutions to the dynamic problems to be considered here. As in the preceding chapters, thermal effects are omitted; they will be included in the more general settings of later chapters.

The main vehicle for our study of one-dimensional dynamics of two-phase materials is the *impact problem*. There is an enormous body of experimental literature pertaining to the response of solids to shock or impact loading, much of it motivated by questions concerning the behavior of materials at extremely high pressures, as occurs, for example, deep in the earth. The reader will find some guidance to the experimental literature in this field of the dynamic behavior of materials in the books by Graham [11] and Meyers [25].

In the prototypical experiment, a specimen of the material under investigation consists of a circular cylindrical disk restrained against radial expansion and subject to normal impact on one of its plane faces by a "flyer-plate" moving at high speed. Care is taken to assure that the plane faces of specimen and flyer-plate are parallel at impact. The collision generates a compressive disturbance that propagates into the disk, ultimately reaching the rear face of the target. Propagation speeds of moving strain discontinuities are measured, as are particle velocities at the rear face. The time scales in such experiments can be as short as tenths of microseconds. Depending on the material, a phase transition may occur, as, for example, in the case of the impact-induced graphite-to-diamond transition studied experimentally by Erskine and Nellis [9].

To begin with, we model the specimen – or target – in an impact experiment as a nonlinearly elastic half-space that is stress free in its initial state, which is taken to be the reference configuration. The flyer-plate is modeled as a massive, rigid half-space that remains in perfect contact with the specimen after striking it. This furnishes the simplest setting in which to exhibit the features of the material model that are associated with phase transitions. To make the connection to experimental observations, however, it will be essential ultimately to replace the half-space target by a slab of infinite lateral extent bounded by parallel planes, since the effect of the specimen's rear face is significant. It is also necessary to account for the elasticity of the flyer-plate if a quantitative comparison is to be made with experimental results. For our purposes, reflections from the circular cylindrical boundary of the target can safely be ignored. It will be assumed that the motion of the half-space is one of uniaxial strain, with each particle moving in the direction normal to the plane boundary. The theory is then one dimensional.

We shall restrict our attention ab initio to a *trilinear* two-phase material. In the theory of quasistatic processes developed in the preceding chapter, no special advantage apart from the simplification of details is offered by the trilinear material. Here, in contrast, it has the major advantage of making possible the construction of simple, explicit solutions to the boundary–initial value problem that models the impact. The special trilinear model to be used undergoes phase transitions in *compression*, as is now physically appropriate.

In the next section, we set out the Lagrangian formulation of the impact problem. At time zero, the particles on the boundary of the half-space are given a prescribed velocity v_0, which is maintained constant for all subsequent times. Since the differential equations, boundary conditions, and initial conditions involve no parameter with units of length or time, it is natural to seek solutions that are invariant under a scale change in space and time. We describe certain general features of such solutions that follow from the dissipation inequality.

In Section 4.3, we exploit the trilinear nature of the material to construct explicit scale-invariant solutions of the impact problem. In the conventional linearized theory of elasticity, the predicted response to the impact loading described above would consist of a single strain discontinuity propagating into undisturbed material, leaving behind it a state of uniform strain, stress, and particle velocity; the speed of propagation of the wave is known a priori. In the *nonlinear* theory of elasticity for a single-phase material with either a strictly convex or a strictly concave stress–strain curve, the response would consist of either a *shock wave* or a *fan*, according as the stress–strain curve in compression is concave or convex; the speed of propagation is *not* known a priori, being determined as part of the solution to the problem. For the trilinear two-phase elastic material, fans cannot occur, and shock waves in either phase travel at speeds that are known in advance. Complications arise solely because of the occurrence of phase boundaries, whose speeds are not only *not* known in advance, but are in fact not determined by the fundamental field equations, boundary conditions, and initial conditions. For weak impacts corresponding to small v_0, we show in Section 4.3 that the solution to the impact problem involves a propagating strain discontinuity separating quiescent material from disturbed material in the same phase. For impacts severe enough to cause a phase transition, the solution may take either of two forms. For impacts of intermediate strength corresponding to moderate values of v_0, the response exhibits a *two-wave* structure: there is a front-running shock wave in the low-pressure phase (LPP), followed by a phase boundary that leaves high-pressure phase (HPP) particles in its wake. There is in fact a one-parameter family of solutions of this kind, in which the parameter is the velocity of the phase boundary. In any one of these solutions, the phase boundary moves subsonically (in a Lagrangian sense) with respect to the velocity of an LPP shock wave. Notably, there is *overlap* between the range of values of v_0 for which the no-phase-change solution exists and the range of v_0 that supports the one-parameter family of solutions with a phase transition. Finally, for the largest values of v_0 corresponding to severe impact, the solution to the impact problem is again uniquely determined by v_0; it has a single-wave structure involving a phase transition. The associated phase boundary moves supersonically with respect to the velocity of an LPP shock wave in this case.

This trichotomy of solution types is present in the responses observed in some experiments; for example, see Erskine and Nellis [9].

The nucleation criterion and the kinetic relation are invoked in Section 4.4. The former chooses between the phase-change and no-phase-change solution in the range of values of v_0 for which both types are available. If this criterion results in a phase transition, the kinetic relation then selects a solution from the one-parameter family of solutions with a phase change. For the severe impacts that arise from large values of v_0, neither the nucleation criterion nor the kinetic relation is needed to determine the supersonic phase boundary speed.

To compare quantitatively the results just described with experiments, it is necessary to modify the model in several respects for the sake of realism. In Section 4.5, we briefly sketch these modifications, none of which affect the trilinear constitutive law, the nucleation criterion, or the kinetic relation. We also describe some results of a detailed comparison of the predictions of the modified model with the experiments of Erskine and Nellis [9].

We shall point out in Section 4.7 a connection between the features of the impact problem just discussed and some well-known and long-standing mathematical issues relating to uniqueness and admissibility of solutions in the theory of quasilinear systems of partial differential equations.

4.2 The impact problem for trilinear two-phase materials

4.2.1 The constitutive law. Consider an elastic body that occupies the half-space $x \geq 0$ in a stress-free reference configuration. We consider motions of the body in which the particle located at x, y, z in the reference configuration is carried at time t to the point $x + u(x, t), y, z$, where $u(x, t)$ is the displacement in the x-direction. The strain γ and the particle velocity v are defined by $\gamma = u_x$ and $v = u_t$, where subscripts indicate partial derivatives. One must have $\gamma > -1$ to assure that the

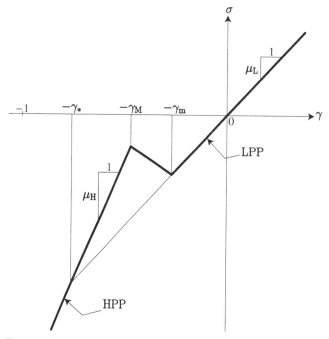

Figure 4.1. The stress–strain curve for a trilinear two-phase elastic material that changes phase in compression between a low-pressure phase (LPP) and a high-pressure phase (HPP).

deformation is one-to-one. The component of nominal (or Piola–Kirchhoff) stress of primary interest here is the normal stress $\sigma(x, t)$ acting on planes perpendicular to the x-axis. In such a uniaxial deformation, there will be other nonvanishing components of stress present as well; we assume that the symmetry of the material is such that no shear stresses arise. This would be the case, for example, if the material were isotropic, though certain forms of anisotropy would also fulfill this requirement, provided the half-space were suitably oriented.

Let $W(\gamma)$ be the strain energy per unit reference volume associated with a uniaxial deformation with strain γ. The stress σ is related to W through $\sigma = W'(\gamma) \equiv \hat{\sigma}(\gamma)$. The stress–strain curve for the *simplest* two-phase material that changes phase in compression has the trilinear form shown in Figure 4.1.

For the trilinear material of Figure 4.1, one has

$$\hat{\sigma}(\gamma) = \begin{cases} \mu_L \gamma, & \text{if} \quad \gamma > -\gamma_m, \\ \mu_H(\gamma + \gamma_T), & \text{if} \quad -1 < \gamma < -\gamma_M. \end{cases} \tag{4.1}$$

The branch of the stress–strain curve for which $\gamma > -\gamma_m$ corresponds to the *low-pressure phase* (LPP) of the material, while the *high-pressure phase* (HPP) refers to the range $-1 < \gamma < -\gamma_M$. The main virtue of the trilinear model is that the nonlinear effects encountered are those associated with the phase transition itself, nonlinearities arising from the bulk behavior of both phases being avoided.

There are two special values of the strain for the trilinear material that will be needed as we proceed. The first of these is $\gamma = -\gamma_*$, the strain at which the

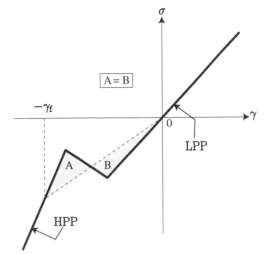

Figure 4.2. The equal-area strain $\gamma = -\gamma_f$ for the trilinear material of Figure 4.1.

extended LPP branch of the stress–strain curve intersects the HPP branch, as shown in Figure 4.1. The second is the strain $\gamma = -\gamma_f$, which, together with $\gamma = 0$, generates a chord that cuts off triangles of equal area from the stress–strain curve; see Figure 4.2. These special strains are given by

$$\gamma_* = \gamma_T/(1 - \mu_L/\mu_H), \quad \gamma_f = \gamma_m + \gamma_M - \gamma_m\gamma_M/\gamma_*. \tag{4.2}$$

We assume that $\gamma_m < \gamma_T < \gamma_M$, so that one has $0 < \gamma_m < \gamma_T < \gamma_M < \gamma_f < \gamma_* < 1$.

In the version of the trilinear material shown in Figure 4.1, the stress $\hat{\sigma}(-\gamma_M)$ at the local maximum has been taken to be compressive (negative), so that the reference state $\gamma = 0$ is the only available stress-free state. If, on the other hand, $\hat{\sigma}(-\gamma_M) > 0$, then there would be a state in the high-pressure phase at which the stress vanishes. In the former case, the impacted body must return to the reference state in the low-pressure phase upon removal of load, whether or not a LPP→HPP phase transition has occurred in response to loading; thus the loading cycle is "globally reversible" in this case. In the latter case $[\hat{\sigma}(-\gamma_M) > 0]$, the body may remain in the high-pressure phase upon unloading if a load-induced phase transition has taken place, in which case the cycle is "globally irreversible." The graphite-to-diamond transition studied in [9], for example, is thought to be globally reversible, corresponding to the case portrayed in Figure 4.1. It is assumed here that $\hat{\sigma}(-\gamma_M) < 0$, as in the figure.

We shall also assume that the high-pressure phase is the stiffer of the two phases, as is often the case for real materials. Thus the moduli for uniaxial deformation in the two phases are assumed to satisfy $\mu_H > \mu_L$. We shall also require that the special strain $-\gamma_*$ be greater than -1 for reasons that will be clear later. These requirements, together with the observation from Figure 4.1 that $\hat{\sigma}(-\gamma_m) < \hat{\sigma}(-\gamma_M)$, lead to the following restriction:

$$(\gamma_M - \gamma_T)/\gamma_m < \mu_L/\mu_H < 1 - \gamma_T. \tag{4.3}$$

When applied to a shock wave in either the low-pressure or the high-pressure phase of the trilinear material, equation (2.22) shows that the speed of propagation of either type of shock is known; the speeds are given by

$$c_L = \sqrt{\mu_L/\rho} \text{ (LPP shock wave)}, \qquad c_H = \sqrt{\mu_H/\rho} \text{ (HPP shock wave)}, \quad (4.4)$$

and $c_L < c_H$; here ρ is the constant referential mass density.

At a shock wave in either phase of the trilinear material, the geometric interpretation of driving force shows immediately that f necessarily vanishes, so that by equation (2.24) the dissipation rate at such a shock wave vanishes as well. This is of course *not* the case for the more general material of Figure 2.2; shock waves in such a more general two-phase material are agents of dissipation, as in gas dynamics.

At a phase boundary with, say, HPP on the left and LPP on the right, one finds from equation (2.25) that, for the trilinear material of Figure 4.1, the driving force is given by

$$f = (\mu_H/2)\left[-\gamma_T \gamma_f - (1 - \mu_L/\mu_H)\gamma^+ \gamma^- - \gamma_T(\gamma^+ + \gamma^-)\right]. \quad (4.5)$$

4.2.2 The impact problem. The target in our impact problem is assumed to be initially at rest in the reference configuration: $v(x, 0) = 0$, $\gamma(x, 0) = 0$. At time $t = 0$, a prescribed constant velocity v_0 is given to the particles at the boundary $x = 0$ of the half-space and maintained for all subsequent times, so that $v(0, t) = v_0$. We wish to construct solutions $\gamma(x, t)$, $v(x, t)$ to the field equations (2.20) that satisfy these boundary and initial conditions and fulfill the jump conditions (2.15) and (2.16), as well as the dissipation inequality (2.26), at any strain discontinuity. Thus we study the following boundary–initial value problem:

$$\left.\begin{aligned}\hat{\sigma}'(\gamma)\gamma_x &= \rho v_t, \\[6pt] v_x &= \gamma_t,\end{aligned}\right\} \quad \text{for } x \geq 0, \quad t \geq 0, \quad (4.6)$$

$$\gamma(x, 0) = v(x, 0) = 0, \quad \text{for } x \geq 0, \quad (4.7)$$

$$v(0, t) = v_0, \quad \text{for } t \geq 0. \quad (4.8)$$

If there is a strain discontinuity at $x = s(t)$, the jump conditions

$$\left.\begin{aligned}[\![\hat{\sigma}(\gamma)]\!] &= -\rho\dot{s}[\![v]\!], \\[6pt] [\![v]\!] &= -\dot{s}[\![\gamma]\!]\end{aligned}\right\} \quad (4.9)$$

and the dissipation inequality

$$f(t)\dot{s}(t) \geq 0 \quad (4.10)$$

must hold. The inequality (4.10) is the counterpart in the present purely mechanical theory of the requirement in adiabatic, inviscid gas dynamics that entropy should increase across a shock wave. For the trilinear material, driving force vanishes at a shock wave in either the LPP or HPP phase, so (4.10) is trivially satisfied at shocks. This is not the case, of course, at phase boundaries.

Because there is no natural length present, the impact problem as just formulated is invariant under the scale change $x \to kx$, $t \to kt$, where $k \neq 0$ is constant; so it is natural to seek solutions that are scale invariant in the sense that $\gamma(kx, kt) = \gamma(x, t)$, $v(kx, kt) = v(x, t)$. One can show that, in such solutions, γ and v must be functions of x/t; *for the trilinear material*, all solutions with this property are piecewise constant, with discontinuities possible only across rays x/t=constant in the x, t-plane; there are no "fans." Thus for a scale-invariant solution of the impact problem, the first quadrant of the x, t-plane may be divided into sectors $S_1, S_2, ..., S_n$ centered at the origin; we take S_1 to be adjacent to the x-axis and S_n adjacent to the t-axis. Thus on S_1, one has $x > \dot{s}_1 t$, and on $S_i : \dot{s}_i t < x < \dot{s}_{i-1} t$ for $i = 2, ..., n$, with $\dot{s}_0 = 0$.

The dissipation inequality (4.10) has significant implications for scale-invariant solutions of the impact problem. First, it allows us to show that a strain corresponding to a point on the unstable, declining branch of the trilinear stress–strain curve cannot arise in any sector S_i. If this occurred, the length of the x-interval in S_i bearing such a strain would increase with time, contradicting an implication of the dissipation inequality established in Section 2.4. Thus no scale-invariant solution of the impact problem can involve states on the declining branch of the trilinear stress–strain curve.

The dissipation inequality (4.10) can also be used to prove that *the solutions we seek can involve no more than one phase boundary and no more than one shock wave*, so that the number n of sectors S_i cannot exceed 3. The proof uses two geometric properties that we observed previously. First, because of (2.22), recall that the speed of a discontinuity is related to the slope of the chord joining the points $(\gamma^-, \hat{\sigma}(\gamma^-))$ and $(\gamma^+, \hat{\sigma}(\gamma^+))$ on the stress–strain curve. Consequently, the speeds of different discontinuities can be compared by comparing the slopes of the corresponding chords. Second, we know from (2.25) that the driving force on a discontinuity can be related to the area between this chord and the stress–strain curve. Figure 4.3 illustrates both of these properties for a phase boundary that has LPP on its right.

We now prove that the number n of sectors S_i cannot exceed 3. To show this, one notes first that on the sector S_1, one will have $\gamma = v = 0$, so that the leading wave front $x = \dot{s}_1 t$ is either (Case 1) a LPP→HPP phase boundary, or (Case 2) a LPP shock wave. Dealing first with Case 1, we show that there can be no discontinuities trailing $x = \dot{s}_1 t$. Suppose the contrary, and let $x = \dot{s}_2 t$, with $\dot{s}_2 < \dot{s}_1$, be the first ray trailing the leading disturbance in the first quadrant of the x, t-plane. This discontinuity must be either a HPP shock wave or a HPP→LPP phase boundary. But sketching the relevant chords on the trilinear stress–strain curve shows immediately that the speed of a HPP shock cannot be less than that of a LPP→HPP phase boundary as required, so $x = \dot{s}_2 t$ must be a HPP→LPP phase boundary. But drawing the chords connecting the relevant states on the stress–strain curve shows that the driving force acting on this phase boundary must be negative if $0 < \dot{s}_2 < \dot{s}_1$, violating the dissipation inequality. Thus if the leading disturbance is a LPP→HPP phase boundary (Case 1), the number of sectors S_i in the first quadrant of the x, t-plane is $n = 2$.

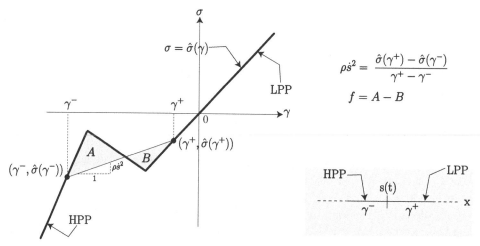

Figure 4.3. A phase boundary with HPP on its left and LPP on its right. The strains on either side of the phase boundary are γ^- and γ^+ respectively. The slope of the chord joining $(\gamma^-, \hat{\sigma}(\gamma^-))$ to $(\gamma^+, \hat{\sigma}(\gamma^+))$ is $\rho\dot{s}^2$ where $\dot{s}$ is the propagation velocity of the phase boundary. The difference in the shaded areas gives the driving force $f = A - B$.

Turning to Case 2, we suppose that the leading disturbance at $x = \dot{s}_1 t$ is a LPP shock wave, so that $\dot{s}_1 = c_L$. As we shall see later, there may or may not be a LPP→HPP phase boundary at $x = \dot{s}_2 t < c_L t$ behind the front-running shock. If there were a still-slower discontinuity at $x = \dot{s}_3 t < \dot{s}_2 t$, it would necessarily be a HPP→LPP phase boundary. For these circumstances, the geometry of the chords connecting the various participating states on the stress–strain curve again leads one to conclude that the driving force at $x = \dot{s}_3 t$ would be negative, violating the dissipation inequality. Thus a LPP shock wave may occur alone or may be followed by a LPP→HPP phase boundary, in which case the number n of sectors S_i in the x, t-plane is either 2 or 3 in Case 2.

The preceding discussion shows that the number n of sectors S_i can be at most 3, and it makes clear that we must consider three possibilities as to solutions of the impact problem: (i) a LPP shock wave alone, corresponding to the absence of a phase change; (ii) a LPP shock wave followed by a LPP→HPP phase boundary; and (iii) a LPP→HPP phase boundary alone. Note that a phase transition occurs in cases (ii) and (iii) but not in (i). Case (iii) involves a phase boundary that necessarily travels at a speed exceeding the shock wave speed, c_L, of LPP; it is often referred to as the *overdriven* case. These three cases will be analyzed in Sections 4.3.1, 4.3.2, and 4.3.3 respectively.

Arguments of the kind just outlined were used in connection with the *Riemann problem* for a different trilinear two-phase material in [1].

4.3 Scale-invariant solutions of the impact problem

4.3.1 Solutions without a phase transition. For sufficiently small values of the impact velocity v_0, one would expect that no phase transition occurs and that the

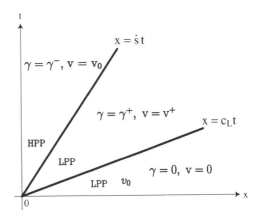

Figure 4.4. Structure in the x, t-plane of a two-wave solution with a LPP→HPP phase transition for the impact problem for the trilinear material. It is necessary that $\gamma^+ > -\gamma_m$ and $-1 < \gamma^- < -\gamma_M$.

solution to the impact problem for the trilinear material would be the conventional one for the linear wave equation corresponding to the low-pressure phase: a LPP shock wave moves into undisturbed material. This solution is easily found to be given by

$$\gamma, \ v = \begin{cases} -v_0/c_L, & v_0 \quad \text{if} \quad 0 < x < c_L t, \\ \\ 0, & 0 \quad \text{if} \quad x > c_L t; \end{cases} \tag{4.11}$$

since this solution assumes the absence of a phase change, it only exists when the strain at each particle remains in LPP at all times, that is, when $-v_0/c_L > -\gamma_m$. This requires that

$$v_0 < c_L \gamma_m. \tag{4.12}$$

4.3.2 Solutions with a phase transition: the two-wave case. For sufficiently large impactor velocity v_0 one would expect a phase change to occur. There are in fact two types of solutions involving a phase transition. In the first type, the target experiences a LPP shock wave moving with speed c_L, followed by a phase boundary; the (Lagrangian) phase boundary velocity $\dot{s}$ must therefore be *subsonic* with respect to the LPP shock speed c_L: $\dot{s} < c_L$. The structure of this "two-wave" solution with a phase transition is indicated in Figure 4.4; the unknowns to be determined are the constant strains $\gamma^\pm$ behind and ahead of the phase boundary, the constant particle velocity v^+ ahead of the phase boundary, and the phase boundary velocity $\dot{s}$.

Since the unknown strains and particle velocities are all constants, the differential equations (4.6) are trivially satisfied where γ, v are smooth. The jump conditions (4.9) then determine $\gamma^\pm$ and v^+ in terms of the impactor velocity v_0 and the *still unknown phase boundary velocity* $\dot{s}$ as follows:

$$\gamma^- = -\frac{(c_L + \dot{s})v_0 + \gamma_T c_H^2}{c_H^2 + c_L \dot{s}}, \quad \gamma^+ = -\frac{(c_H^2 - \dot{s}^2)v_0 - \gamma_T c_H^2 \dot{s}}{(c_L - \dot{s})(c_H^2 + c_L \dot{s})}, \quad v^+ = -c_L \gamma^+.$$
$$\tag{4.13}$$

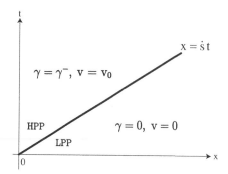

Figure 4.5. Structure in the x, t-plane of a one-wave solution with a LPP→HPP phase transition for the impact problem for the trilinear material. It is necessary that $\gamma^- < -\gamma_M$.

As depicted in Figure 4.4, this solution presumes that particles in the range $\dot{s}t < x < c_L t$ are in LPP, while those in $0 < x < \dot{s}t$ are in HPP; consequently, the corresponding strains γ^+ and γ^- are required to be such that

$$\gamma^+ > -\gamma_m, \qquad -1 < \gamma^- < -\gamma_M. \tag{4.14}$$

Through (4.13), the velocities v_0 and $\dot{s}$ are subject to restrictions arising from the phase segregation inequalities (4.14). From (4.5) and (4.13), one can find the driving force at the phase boundary in terms of the given impactor velocity v_0 and the as yet unknown phase boundary velocity $\dot{s}$. The result may be written in the form

$$f = -P\left(\frac{v_0}{\gamma_T c_H}\right)^2 + 2Q\left(\frac{v_0}{\gamma_T c_H}\right) - R, \tag{4.15}$$

where P, Q, and R are functions of $\dot{s}$ defined by

$$P = \mu_H(\gamma_T^2/2)(c_H{}^2 - c_L{}^2)\frac{c_H{}^2 - \dot{s}^2}{(c_H{}^2 + c_L\dot{s})^2}\frac{c_L + \dot{s}}{c_L - \dot{s}}, \tag{4.16}$$

$$Q = \mu_H(\gamma_T^2/2)c_L c_H \frac{c_H{}^2 - \dot{s}^2}{(c_H{}^2 + c_L\dot{s})^2}\frac{c_L + \dot{s}}{c_L - \dot{s}}, \tag{4.17}$$

$$R = \mu_H(\gamma_T^2/2)\left\{\frac{\gamma_f}{\gamma_T} + \frac{2c_L c_H{}^2\dot{s}^2 - c_H{}^4(c_L - \dot{s})}{(c_H{}^2 + c_L\dot{s})^2(c_L - \dot{s})}\right\}. \tag{4.18}$$

Through (4.15), a further restriction – to be discussed shortly – is imposed on v_0 and $\dot{s}$ by the dissipation inequality (4.10).

4.3.3 Solutions with a phase transition: the one-wave case. The second type of
solution involving a phase change has the structure shown in Figure 4.5.

In this case, each particle of the target jumps directly from the undisturbed LPP reference state to the HPP state; the phase boundary is not preceded by a shock wave. In contrast to the two-wave case just discussed, applying the jump conditions at the phase boundary now *fully determines the unknowns γ^- and $\dot{s}$ in terms of the*

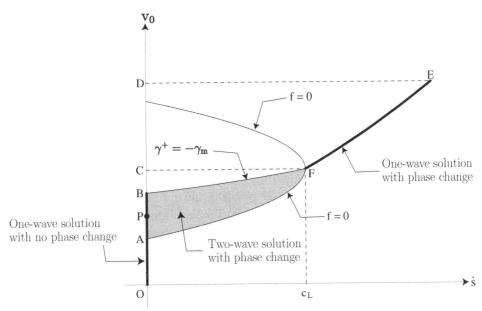

Figure 4.6. Pairs $(\dot{s}, v_0)$ for which the impact problem has a solution. The segment OB of the vertical axis shown bold, defined by (4.12), corresponds to solutions without a phase transition. The shaded region is defined by (4.13), (4.14), and driving force $f > 0$; it corresponds to two-wave solutions with a phase transition. One-wave solutions with a phase transition are associated with points on the curve FE, which is defined by (4.19) and (4.21).

impactor velocity v_0:

$$\gamma^- = -(\gamma_T/2)\left\{1 + \sqrt{1 + 4v_0^2/(c_H^2\gamma_T^2)}\right\}, \qquad \dot{s} = \frac{2v_0/\gamma_T}{1 + \sqrt{1 + 4v_0^2/(c_H^2\gamma_T^2)}}.$$
$$(4.19)$$

Through $(4.19)_1$, v_0 must be restricted by the phase segregation condition $-1 < \gamma^- < -\gamma_M$.

Using the representation (4.5) with $\gamma^+ = 0$ and γ^- given by (4.19), one can find the driving force acting on the phase boundary for the one-wave solution. The requirement imposed by the dissipation inequality (4.10) then becomes a further restriction on γ^-; this restriction turns out to be $\gamma^- < -\gamma_f$, where γ_f is the special equal-area strain introduced earlier in (4.2) and Figure 4.2. In addition, γ^- must satisfy $-1 < \gamma^- < -\gamma_M$. Since $\gamma_f > \gamma_M$, these requirements reduce to

$$-1 < \gamma^- < -\gamma_f, \qquad (4.20)$$

which, through $(4.19)_1$, becomes the following restriction on the impactor velocity:

$$\sqrt{\gamma_f(\gamma_f - \gamma_T)}\, c_H < v_0 < \sqrt{1 - \gamma_T}\, c_H. \qquad (4.21)$$

4.3.4 The totality of solutions. It is helpful to describe the solutions discussed above in a Cartesian plane in which the coordinates are the phase boundary velocity $\dot{s}$ and the impactor velocity v_0, as in Figure 4.6.

We first delineate the set of points in this plane that correspond to the no-phase-change solution (4.11) permitted by the restriction (4.12). Thinking of these formally as special cases of the two-wave solutions of Figure 4.4 for which $\dot{s} = 0$, we may say that each no-phase-change solution corresponds to a point in the interval $0 < v_0 < c_L\gamma_m$ of the vertical axis shown bold in Figure 4.6.

Next we describe the points $(\dot{s}, v_0)$ corresponding to the two-wave solutions (4.13) *with* a phase change, subject to the phase segregation conditions (4.14) and the dissipation inequality (4.10), with f given by (4.5). It turns out that, of these inequalities, the decisive ones are $\gamma^+ > -\gamma_m$ and $f > 0$; all others are implied by these. Using (4.5), one finds that $f = 0$ corresponds to the parabola-like curve shown schematically in Figure 4.6; for points inside this curve, one has $f > 0$. The requirement $\gamma^+ > -\gamma_m$ corresponds to points below the curve marked $\gamma^+ = -\gamma_m$, also sketched schematically in the figure. Thus each point in the shaded region of Figure 4.6 corresponds through (4.13) to a two-wave solution satisfying all requirements.

Finally, the points in the $\dot{s}$, v_0-plane that correspond to the one-wave, phase-changing solutions lie on the curve represented by the second of (4.19), subject to the restriction (4.21). A portion of this curve, not shown in Figure 4.6, lies *inside* the shaded region in the figure, while the remainder lies outside this region and is shown bold as the curve FE. The points on the curve that lie inside the shaded area correspond to special cases of the *two*-wave solution (4.13) for which the front-running shock wave happens to be absent whence $\gamma^+ = 0$. For this to occur, $\dot{s}$ and v_0 must be suitably related; the necessary relation is precisely the second of (4.19). Thus only the one-wave solutions associated with points on FE correspond to solutions not already accounted for in (4.13).

The following conclusions can be drawn from Figure 4.6:

(*i*) *For each impactor velocity v_0 in the interval OA of the vertical axis, there is a unique solution given by (4.11); there is no phase transition.*

(*ii*) *For each v_0 corresponding to a point, say P, between A and B on the vertical axis, there are two types of solution: a no-phase-change solution (4.11) corresponding to the point P itself, and a one-parameter family of two-wave solutions (4.13) corresponding to all points in the shaded region that lie on the horizontal line through P. Each of the latter describes a phase transition.*

(*iii*) *For each v_0 corresponding to a point on the vertical axis between B and C, there is a one-parameter family of two-wave solutions (4.13) in each of which the target ultimately changes phase. Solutions without a phase change cannot occur.*

(*iv*) *For each v_0 corresponding to a point on the vertical axis between C and D, there is a unique solution; it is the one-wave solution (4.19) with a phase transition.*

(*v*) *For each v_0 corresponding to a point on the vertical axis above D, there is no solution to the impact problem.*

According to (i) above, unless the impactor velocity is large enough, there will be no phase transition.

Conclusion (ii) describes the way in which the impact problem differs drastically from its counterpart for a single-phase material: there is a massive loss of uniqueness. This loss manifests itself in two ways; for impactor velocities v_0 in the segment AB of the v_0-axis, the target must first "choose" between the no-phase-change solution and the one-parameter family (parameter $\dot{s}$) of two-wave, phase-changing solutions that are *also* available for such a v_0. As we shall see below, this choice is made by a *nucleation criterion*. If the choice favors the phase transition, the body must then select a *particular* two-wave solution from the family (4.13). This is done through a *kinetic relation*, which – as we show below – determines $\dot{s}$ in terms of v_0, thus selecting a solution from the one-parameter family.

If v_0 corresponds to a point on the segment BC, conclusion (iii) shows that infinitely many solutions are again available, all involving a phase transition; a kinetic relation is again needed to choose a particular one.

If v_0 is large enough to fall in the segment CD of the vertical axis, conclusion (iv) asserts that there is a unique one-wave solution (4.19), and it involves a phase transition. A kinetic relation is not needed and may not be prescribed. It may be noted that the speed $\dot{s}$ of the phase boundary in this case exceeds the speed c_L of LPP shock waves; the phase boundary is said to be *overdriven*. The distinction between the two-wave solution, for which a kinetic relation must be prescribed, and the one-wave solution, where kinetics may *not* be imposed, apparently has a parallel in combustion theory, where *deflagrations* require prescription of the relevant kinetics, while *detonations* do not; see [14].

Finally, the fact that there is *no* solution for all sufficiently large values of v_0, as asserted in conclusion (v), is a consequence of the fact that, according to the predictions of the model, such impacts result in strains of magnitude greater than unity, corresponding to infinite contraction of the target at a finite stress. This is an artifact of the trilinear model; it would not occur for a two-phase material model in which the stress $\hat{\sigma}(\gamma)$ in the high-pressure phase is unbounded as the strain γ tends to -1.

4.4 Nucleation and kinetics

As in the case of quasistatic processes discussed in the preceding chapter, we resolve the issue of loss of uniqueness of solution of the impact problem by appealing to a nucleation criterion and a kinetic relation. We postulate that the nucleation site is the plane boundary of the half-space where the impact occurs. Since in uniaxial deformation, all particles in this plane are in the same mechanical state at every instant, nucleation will cause all of them to undergo the LPP→HPP transition simultaneously, causing a plane phase boundary to emerge from the site and propagate into the interior of the body. As before, LPP→HPP nucleation occurs if the driving force it would generate at the incipient phase boundary is at least as large as the materially determined critical value f_n^{LH}.

Once a phase boundary emerges, its propagation is controlled by the kinetic relation. As in the quasistatic case, it is natural to assume that the Lagrangian propagation speed $\dot{s}$, which represents the rate at which material transforms from one phase to the other, is determined by the local states on either side of the phase boundary, that is, that it is determined by γ^+ and γ^-. Accordingly we assume that there is a kinetic relation of the form

$$\dot{s} = \tilde{v}(\gamma^+, \gamma^-). \tag{4.22}$$

Next, applying equation (2.22) to a phase boundary with HPP on the left and using the trilinear stress–strain relation (4.1) leads to

$$\dot{s}^2 = \frac{c_L^2 \gamma^+ - c_H^2(\gamma^- + \gamma_T)}{\gamma^+ - \gamma^-}. \tag{4.23}$$

The respective equations (4.23) and (4.5) provide expressions for the phase boundary velocity $\dot{s}$ and driving force f in terms of the strains $\gamma^{\pm}$ on either side of the phase boundary. Solving this pair of equations for $\gamma^{\pm}$ allows us to express the strains γ^+ and γ^- in terms of f and $\dot{s}$. The kinetic relation $\dot{s} = \tilde{v}(\gamma^+, \gamma^-)$ can therefore be written in the equivalent form $\dot{s} = \bar{v}(f, \dot{s})$. We shall restrict our attention to the special case of this,

$$\dot{s} = \Phi(f) \quad \text{or} \quad f = \phi(\dot{s}). \tag{4.24}$$

Returning to the impact problem, once the phase boundary emerges from the impact plane, its propagation is controlled by the kinetic relation (4.24) where the driving force f is given by (4.5), as is appropriate for a phase boundary that has HPP on its left. Recalling that, for the one-parameter family of two-wave solutions (4.13), the phase boundary moves subsonically with respect to LPP shock waves, it is assumed that $\phi(\dot{s})$ is defined for $-c_L < \dot{s} < c_L$, is monotonically nondecreasing there, is continuous except possibly at $\dot{s} = 0$, and satisfies $\phi(\dot{s})\dot{s} \geq 0$, as required by the dissipation inequality (4.10). A typical kinetic response function is shown in Figure 3.4.

With $\phi(\dot{s})$ given, one substitutes for f on the left side of $(4.24)_2$ the expression in terms of v_0 and $\dot{s}$ given in (4.15) for the driving force acting on the phase boundary in the two-wave solution. The result is a relation between the impactor velocity and the phase boundary velocity with which to find $\dot{s}$ in terms of v_0 and hence to fully determine the two-wave solution (4.13).

To illustrate the procedure just described, we consider a special kinetic relation, termed *maximally dissipative* in [2], which results when $\phi(\dot{s})$ is chosen so as to maximize the dissipation rate for each fixed $\dot{s}$. The notion of maximum dissipation intended here is analogous to that of *maximum plastic work* used in the constitutive description of rate-independent elastic–plastic solids; see, for example, [24, 31]. To determine such a $\phi(\dot{s})$, one uses the representation (4.5) for f to maximize $f\dot{s}$ with respect to either γ^+ or γ^-, subject to the constraint (4.23), *with $\dot{s}$ fixed*. Although $\dot{s}$ is necessarily positive in the impact problem, we temporarily permit negative values of $\dot{s}$ corresponding to the HPP→LPP transition in order to fully exhibit the maximally dissipative kinetic relation. For $\dot{s} > 0$, one finds that the

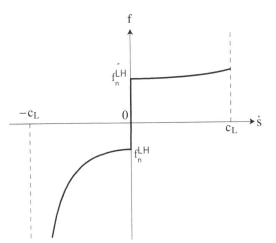

Figure 4.7. Schematic graph of the relation between driving force f and phase boundary velocity $\dot{s}$ for the maximally dissipative kinetic relation (4.25).

maximizing value of γ^+ is given by $\gamma^+ = -\gamma_m$, while if $\dot{s} < 0$, the maximizing value of $\gamma^- = -\gamma_M$. This leads through (4.5) and (4.23) to the following special case of $(4.24)_2$:

$$f = \phi(\dot{s}) = \frac{\mu_H - \mu_L}{2}\left\{\gamma_*(\gamma_* - \gamma_f) - \phi_0(\dot{s})\right\}, \qquad (4.25)$$

where

$$\phi_0(\dot{s}) = \begin{cases} (\gamma_* - \gamma_M)^2(c_H{}^2 - \dot{s}^2)/(c_L{}^2 - \dot{s}^2), & \text{if} \quad -c_L < \dot{s} < 0, \\ \\ (\gamma_* - \gamma_m)^2(c_L{}^2 - \dot{s}^2)/(c_H{}^2 - \dot{s}^2), & \text{if} \quad 0 < \dot{s} < c_L. \end{cases} \qquad (4.26)$$

For maximally dissipative kinetics, the critical value of f for LPP$\to$HPP nucleation is chosen to be $f_n^{LH} = \phi(0+)$, which is also the resistance f_r^{LH} to LPP$\to$HPP phase boundary motion. A schematic graph of (4.25) is shown in Figure 4.7.

Since $\gamma^+ = -\gamma_m$ for maximally dissipative kinetics in the case of positive $\dot{s}$ of interest to us in the impact problem, reference to the stress–strain curve of Figure 4.1 suggests the following interpretation of the process just described: the phase change is not initiated until the given impact velocity is sufficient to drive the magnitude of the compressive strain in the target to its largest possible value in the low-pressure phase, corresponding to the local minimum at $\gamma = -\gamma_m$ in Figure 4.1. Thus the particles at which nucleation is to take place "wait as long as possible" to jump to the new phase, corresponding to a convention sometimes called *perfect delay* [30].

It is clear that (4.15)–(4.18), (4.25), and (4.26) together provide a relation between impactor velocity v_0 and phase boundary velocity $\dot{s}$. This relation is most efficiently obtained by setting $\gamma^+ = -\gamma_m$ in $(4.13)_2$; the result is

$$v_0 = \gamma_m c_L + \left\{\gamma_T c_H{}^2 - \gamma_m(c_H{}^2 - c_L{}^2)\right\}\frac{\dot{s}}{c_H{}^2 - \dot{s}^2}, \qquad (4.27)$$

$$0 \le \dot{s} < c_L \text{ (two-wave response)}.$$

The right side of (4.27) is an increasing, and thus invertible, function of $\dot{s}$.

For the overdriven one-wave solution, the counterpart of (4.27) is (4.19)$_2$, which may be inverted to give

$$v_0 = \gamma_T c_H^2 \frac{\dot{s}}{c_H^2 - \dot{s}^2}, \quad c_L < \dot{s} < \sqrt{1 - \gamma_T} c_H \text{ (one-wave response).} \quad (4.28)$$

The time history of particle velocity is of interest in experiments. Assuming that a phase change occurs and that maximally dissipative kinetics apply, one finds $v(x, t)$ as a function of t at fixed x from (4.13)$_3$ and the fact that $\gamma^+ = -\gamma_m$:

$$v(x, t) = \begin{cases} 0, & \text{for} \quad 0 < t < x/c_L \\[2mm] \gamma_m c_L, & \text{for} \quad x/c_L < t < x/\dot{s} \quad \text{(two-wave response),} \\[2mm] v_0, & \text{for} \quad t > x/\dot{s} \end{cases} \quad (4.29)$$

where the phase boundary speed $\dot{s}$ is related to the impactor velocity v_0 by (4.27). It may be noted from (4.29) that the particle velocity behind the front-running shock wave is independent of the impactor velocity. As we shall see below, this artifact of maximally dissipative kinetics is observed, at least approximately, in some experiments.

If the impactor velocity is great enough to cause overdriven response, one has instead the simple relation

$$v(x, t) = \begin{cases} 0, & \text{for} \quad 0 < t < x/\dot{s} \\[2mm] v_0, & \text{for} \quad t > x/\dot{s} \end{cases} \quad \text{(one-wave response),} \quad (4.30)$$

where now $\dot{s}$ is determined in terms of v_0 by (4.28).

4.5 Comparison with experiment

Impact experiments usually involve cylindrical specimens, one of whose plane faces is struck by a flyer-plate traveling perhaps several kilometers per second. Particle velocities are measured near the center of the face of the specimen opposite to the impact face, often by passing a laser beam through an optically clear material, or "window," in contact with the specimen at its rear face, exploiting the Doppler effect. A clear description of such experiments involving a shock-induced graphite-to-diamond phase transition may be found, for example, in [9].

Figure 4.8 is taken from [9]; it shows the particle velocity histories at the rear face of the graphite specimen observed in four experiments corresponding to different impactor velocities. The highest of the four impactor velocities produces one-wave, overdriven response; the other three traces show the two-wave structure. Measurements were made of the speeds of propagation of both first and second wave fronts as well as the particle velocities behind them.

The experimental results in [9] exhibit three features that are consistent with the trilinear material model accompanied by maximally dissipative kinetics. First,

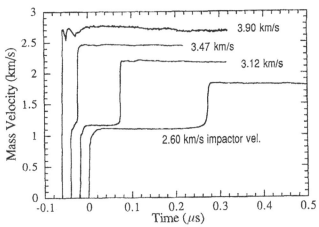

Figure 4.8. Particle velocity versus time as measured in the experiments of Erskine and Nellis [9]. The target was pyrolytic (ZYB) graphite. The curves have been staggered horizontally on the graph for clarity. The two-wave structure is direct evidence for a phase transition. Reprinted from [9] with permission from the Journal of Applied Physics.

the observed history of particle velocity has the two-wave structure predicted by the model discussed above for all but the highest impactor velocity achieved in the experiments. For this most severe impact, the experimental result has the one-wave, overdriven structure that also arises in our analysis. Second, the velocity of the front-running shock wave observed in the two-wave experiments is nearly independent of impactor velocity, being almost constant over the range of impactor velocities that result in two-wave response. This is consistent with the fact that, in the trilinear material, the front-running shock travels with constant velocity c_L. Third, in the experimentally observed two-wave particle velocity histories, the particle velocities observed after the arrival of the front-running shock but prior to the arrival of the phase boundary are *also* independent of impactor velocity to a good degree of approximation. Thus the experimental results in [9] are consistent in this sense with the predictions for the impact problem of the model based on the trilinear material and maximally dissipative kinetics.

The preceding analysis can be applied to the experiments described in [9], but only after several essential modifications. One must account for the deformability of the flyer-plate, tacitly assumed to be rigid in the preceding analysis, as well as the presence of the deformable window in contact with the specimen at its rear face. The latter modification requires that the "half-space target" in the analysis above be replaced by a slab of finite thickness, a change that necessitates the consideration of waves reflected and transmitted at the target-window interface. An analysis incorporating these modifications was carried out in [4]. The experimentally determined propagation speeds of the two waves and the observed particle velocities as given in [9] were used to determine the parameters γ_m, γ_T, c_L, and c_H. These values, together with the density ρ, allow the calculation of the particle velocities as functions of time as predicted by the model for the impactor velocities arising in [9]. The results are shown in Figure 4.9; the graphs are the

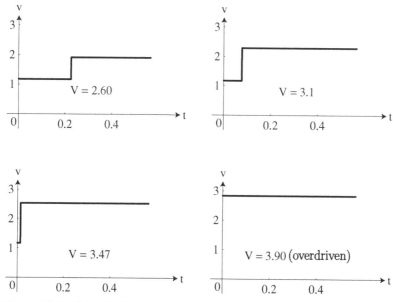

Figure 4.9. Particle velocities versus time as predicted by the model in [4] for impactor velocities in the experiments of [9]. Graphs correspond to the experimental results shown in Figure 4.8. Units on the time axis are microseconds; those on the particle velocity axis are kilometers per second. Time zero is taken as the arrival time of the first wave. As in the experiments, the predicted response at the highest impactor velocity is overdriven.

counterparts for the model of those for the experiments shown in Figure 4.8. Table 4.1 compares the numerical values of the ultimate particle velocity v_{model} as predicted by the model with the observed values v_{exp}. The largest error between predicted and observed values is about 5.7% of the observed value. For further details comparing the model with the experiments reported in [9], the reader is referred to [4].

Table 4.1. Predicted values of ultimate particle velocity v_{model} and observed values v_{exp} for various impactor velocities v_0 in the graphite-to-diamond phase transitions. Experimental values are taken from Erskine and Nellis [9]; predicted values are as given in [4]. All velocities are in kilometers per second. Reproduced with permission from Journal of Applied Physics.

v_0	v_{model}	v_{exp}
2.60	1.91	1.83
3.12	2.28	2.20
3.47	2.53	2.46
3.90[a]	2.84[a]	2.70[a]

[a] Overdriven.

4.6 Other types of kinetic relations

A kinetic relation relates the rate at which the material transforms from one phase to the other to the local states of the material immediately before and after transformation. Given a particular transformation, the associated kinetic relation can be derived, in principle, if the dynamics of the phase transition at the microscopic scale can be suitably modeled. In effect, the role of a kinetic relation is to provide a continuum scale characterization of the essential features of the physics of the microscopic transformation mechanisms. Several examples of kinetic relations based on various physical mechanisms will be described in Section 8.4.

On the other hand the kinetic relation given by (4.25) and (4.26) arises from the phenomenological assumption of maximal dissipation. Another instance of a phenomenologically based kinetic relation is that for dissipation-free phase boundary motion: $f = \phi(\dot{s}) = 0$.

In the setting of the impact problem discussed earlier in this chapter, the kinetic relation picks a unique solution from a one-parameter family of available solutions that describe the impact-induced phase transition and involve a propagating sharp interface separating the low- and high-pressure phases of the material. In this sense, a kinetic relation serves also as a *selection principle* to be applied when the conventional formulation of the impact problem leads to a breakdown in uniqueness of solution.

Another example of such a selection principle, which for our basic system (4.6) is also equivalent to a phenomenologically based kinetic relation of the form (4.24), is the *entropy rate admissibility criterion* of Dafermos [6, 7]. This criterion maximizes the dissipation rate in a sense different from that employed earlier in this chapter; for a detailed discussion of the two notions of maximum dissipation rate, the reader is referred to [2]. The entropy-rate admissibility condition has been used in the setting of phase transitions by Hattori [12, 13], James [15], and Pence [29].

Though the function $\phi(\dot{s})$ in the kinetic relation $f = \phi(\dot{s})$ is usually taken to be monotonically increasing, the possibility of nonmonotonic kinetic response functions has been explored in [32]. For a trilinear material, the analysis in [32] addresses the stability of the impact problem under a perturbation of the boundary data in the presence of nonmonotonic kinetics; it is found that such a perturbation may lead to stick-slip phase boundary motion, a phenomenon sometimes seen in experiments.

As mentioned in Section 3.4, the effect of nonhomogeneous kinetics, in which the driving force f depends on the instantaneous position $x = s$ as well as the velocity $\dot{s}$ of the phase boundary, has been addressed in the setting of quasistatic processes. We know of no attempt to study such effects under truly dynamic conditions.

4.7 Related work

Continuum models of the macroscopic response of solids in the present purely mechanical context, with thermal effects omitted, have been applied to particular physical problems by various authors. Thus Zhong [38] has discussed the effect of an elastic wave incident upon the interface between two phases of a trilinear elastic bar; he finds the long-time limit of the position of the phase boundary after it is dislodged from its initial position by the incident wave.

The analysis in [3] concerns a composite bar, half of which is made of a single-phase elastic material while the remaining half is trilinear. A wave traveling in the single-phase material is incident on the bond between the two halves of the bar. The question addressed is whether the phase-changing capability of the trilinear material can be exploited to augment or diminish the strength of the wave that is either reflected or transmitted from the bond.

Many studies of one-dimensional dynamic problems for phase-changing elastic bodies have been limited to trilinear materials. In [22], Lin analyzes Riemann problems for a two-phase elastic material whose *bulk* behavior in either phase is nonlinear, so that the relevant stress–strain curve looks like that in Figure 2.2.

A computational scheme capable of coping with propagating phase boundaries has been developed by Zhong, Hou, and LeFloch [39]; this scheme has been generalized and applied to the interpretation of impact experiments by Winfree [37].

The second-order, quasilinear system (4.6) for the two unknowns γ, v has been the subject of extensive study, perhaps motivated initially by interest in gas dynamics [5], but often simply as a system of mathematical interest, with pioneering contributions from Dafermos [7], Lax [18], Liu [23], Oleinik [26], and others. For general expositions of the theory, the reader may consult the notes by Lax [19], the textbook of Evans [10], or the recent monograph of LeFloch [21], which emphasizes quasilinear systems to which supplementary restrictions such as kinetic relations must be appended.

In the purely mechanical setting of the present chapter, the dissipation inequality (4.10) for the system (4.6) is the counterpart of the entropy inequality arising from the second law of thermodynamics; its role in selecting a unique weak solution to problems in gas dynamics is well known. Many authors have studied admissibility criteria intended to play the role for general systems of conservation laws that the entropy inequality plays in gas dynamics; see [21] for references. One of the most widely used such criteria involves replacing the nonlinear constitutive law by one that adds effects due to viscosity and strain gradient to the elastic part of the stress:

$$\sigma = \hat{\sigma}(\gamma) + v\gamma_t - \lambda\gamma_{xx}. \tag{4.31}$$

Viscosity–strain gradient criterion:
For the system (4.6), a strain discontinuity propagating with velocity $\dot{s}$ is admissible if it is the limit as v and λ tend to zero of a traveling wave with velocity $\dot{s}$ for the system obtained from momentum balance, compatibility, and (4.31).

The use of this criterion in the study of phase transitions for the system (4.6) was pioneered by Slemrod [34] and Truskinovsky [36]. Another important criterion was proposed by Lax [18, 19]. For the system (4.6), the Lax admissibility criterion is equivalent to the following condition given by Dafermos [7].

Lax criterion:
A strain discontinuity propagating with velocity $\dot{s}$ and joining the states γ^+, v^+ and γ^-, v^- is admissible if $\dot{s}(\gamma^+ - \gamma^-) \geq 0$ ($\dot{s}(\gamma^+ - \gamma^-) \leq 0$) and the chord joining the points $(\gamma^+, \hat{\sigma}(\gamma^+))$ and $(\gamma^-, \hat{\sigma}(\gamma^-))$ on the stress–strain curve lies below (above) the tangents at $(\gamma^+, \hat{\sigma}(\gamma^+))$ and $(\gamma^-, \hat{\sigma}(\gamma^-))$.

For nonlinearly elastic materials for which $\hat{\sigma}(\gamma)$ is either a strictly convex or strictly concave function of γ – the so-called *genuinely nonlinear* case [19] – the geometric interpretation of driving force allows one to show that the dissipation inequality (4.10) is equivalent to the Lax criterion. It is known that imposing the Lax criterion at all strain discontinuities is sufficient in the genuinely nonlinear case for the uniqueness of the solution to the Riemann problem associated with the system (4.6); see Chapter VI of [21]. On the other hand, for materials that are *not* genuinely nonlinear such as the ones with which we are concerned here, the Riemann problem need not have a unique solution, even with the Lax condition in force; see [21] as well as the remarks of Dafermos [8].

Still another admissibility condition, the so-called *chord criterion*, has been studied by various authors after being proposed by Oleinik [27] for first-order scalar conservation laws. As stated by Shearer [33], the chord criterion would apply in our setting to *any* stress–strain relation $\sigma = \hat{\sigma}(\gamma)$.

Chord criterion:
A strain discontinuity propagating with velocity $\dot{s}$ and joining the states γ^+, v^+ and γ^-, v^- is admissible if either $\dot{s} = 0$, or $\dot{s}(\gamma^+ - \gamma^-) \geq 0$ ($\dot{s}(\gamma^+ - \gamma^-) \leq 0$) and the chord joining the points $(\gamma^+, \hat{\sigma}(\gamma^+))$ and $(\gamma^-, \hat{\sigma}(\gamma^-))$ on the stress–strain curve lies below (above) the curve.

The geometric interpretation of the driving force allows one to show that any discontinuity that is admissible according to the chord criterion also satisfies the dissipation inequality (4.10), though the converse is false. In the genuinely non-linear case, the chord criterion and the Lax criterion are equivalent. Pego [28] has shown that, under suitable assumptions, the chord criterion is equivalent to a *viscosity criterion* obtained as a special case of the viscosity–strain gradient criterion mentioned above by choosing the strain gradient coefficient λ to be zero in (4.31).

For two-phase elastic materials, the stress–strain relation $\sigma = \hat{\sigma}(\gamma)$ not only fails to be genuinely nonlinear, it fails to be monotonically increasing as well; the system (4.6) is then hyperbolic-elliptic. For an elastic material like rubber, the curvature of the stress–strain curve changes from concave to convex as the tensile strain increases [35], so that genuine nonlinearity fails. Monotonicity holds, however, so that the system (4.6) remains hyperbolic in this case. For a rubberlike bar, a kinetic relation turns out to be needed in the impact problem; see [16, 17]. Hyperbolic systems that fail to be genuinely nonlinear are discussed in [21], where references to other work in this area may be found.

An existence theorem in the dynamics of two-phase elastic materials was established by LeFloch [20].

REFERENCES

[1] R. Abeyaratne and J.K. Knowles, Kinetic relations and the propagation of phase boundaries in solids. *Archive for Rational Mechanics and Analysis*, **114** (1991), pp. 119–54.

[2] R. Abeyaratne and J.K. Knowles, On the propagation of maximally dissipative phase boundaries in solids. *Quarterly of Applied Mathematics*, **50** (1992), pp. 149–72.

[3] R. Abeyaratne and J.K. Knowles, Reflection and transmission of waves from an interface with a phase-transforming solid. *Journal of Intelligent Materials Systems and Structures*, **3** (1992), pp. 224–44.

[4] R. Abeyaratne and J.K. Knowles, On a shock-induced martensitic phase transition. *Journal of Applied Physics*, **87** (2000), pp. 1123–34.

[5] R. Courant and K.O. Friedrichs, *Supersonic Flow and Shock Waves*, Interscience. New York, 1948.

[6] C. Dafermos, The entropy rate admissibility criterion for solutions of hyperbolic conservation laws. *Journal of Differential Equations*, **14** (1973), pp. 202–12.

[7] C. Dafermos, Hyperbolic systems of conservation laws, in *Systems of Nonlinear Partial Differential Equations*, edited by J.M. Ball. Reidel, Dordrecht, 1983, pp. 25–70.

[8] C. Dafermos, Discontinuous thermokinetic processes, Appendix 4B. *Rational Thermodynamics*, edited by C. Truesdell. Springer-Verlag, New York, 1984.

[9] D.J. Erskine and W.J. Nellis, Shock-induced martensitic transformation of highly oriented graphite to diamond. *Journal of Applied Physics*, **71** (1992), pp. 4882–6.

[10] L.C. Evans, *Partial Differential Equations*. American Mathematical Society, Providence, RI, 1998.

[11] R.A. Graham, *Solids under High-Pressure Shock Compression*. Springer-Verlag, New York, 1992.

[12] H. Hattori, The Riemann problem for a van der Waals fluid with entropy rate admissibility criterion. Isothermal case. *Archive for Rational Mechanics and Analysis*, **92** (1986), pp. 247–63.

[13] H. Hattori, The Riemann problem for a van der Waals fluid. Nonisothermal case. *Journal of Differential Equations*, **65** (1986), pp. 158–74.

[14] W.D. Hayes, *Gas Dynamic Discontinuities*. Princeton University Press, Princeton, 1960.

[15] R.D. James, The propagation of phase boundaries in elastic bars. *Archive for Rational Mechanics and Analysis*, **73** (1980), pp. 125–58.

[16] J.K. Knowles, Impact-induced tensile waves in a rubberlike material. *SIAM Journal on Applied Mathematics*, **62** (2002), pp. 1153–75.

[17] J.K. Knowles, Sudden tensile loading of a rubberlike bar. *Mechanics Research Communications*, **30** (2003), pp. 581–7.

[18] P. Lax, Hyperbolic systems of conservation laws II. *Communications on Pure and Applied Mathematics*, **10** (1957), pp. 537–66.

[19] P. Lax, *Hyperbolic Systems of Conservation Laws and the Mathematical Theory of Shock Waves*, Regional Conference Series in Applied Mathematics, No. 11, SIAM, Philadelphia, PA, 1973.

[20] P.G. LeFloch, Propagating phase boundaries: formulation of the problem and existence via the Glimm method. *Archive for Rational Mechanics and Analysis*, **123** (1993), pp. 153–97.

[21] P.G. LeFloch, *Hyperbolic Systems of Conservation laws*. Birkhäuser Verlag, Basel, 2002.

[22] Y. Lin, A Riemann problem for an elastic bar that changes phase. *Quarterly of Applied Mathematics*, **53** (1995), pp. 575–600.

[23] T.P. Liu, The Riemann problem for general 2×2 conservation laws. *Transactions of the American Mathematical Society*, **199** (1974), pp. 89–112.

[24] J. Lubliner, A maximum dissipation principle in generalized plasticity. *Acta Mechanica*, **52** (1984), pp. 225–37.

[25] M. Meyers, *Dynamic Behavior of Materials*. Wiley, New York, 1994.

[26] O. Oleinik, On the uniqueness of the generalized solution of the Cauchy problem for a nonlinear system of equations occurring in mechanics. *Uspekhi Matematicheskii Nauk (N.S.)*, **12** (1957), pp. 169–76 (in Russian).

[27] O. Oleinik, Uniqueness and stability of the generalized solution of the Cauchy problem for a quasilinear equation. *Uspekhi Matematicheskii Nauk (N.S.)*, **14** (1959), pp. 165–70 (in Russian).

[28] R. Pego, Phase transitions in one-dimensional nonlinear viscoelasticity: admissibility and stability. *Archive for Rational Mechanics and Analysis*, **97** (1986), pp. 353–94.

[29] T.J. Pence, On the encounter of an acoustic shear pulse with a phase boundary in an elastic material: energy and dissipation. *Journal of Elasticity*, **25** (1991), pp. 31–74.

[30] T. Poston and I.N. Stewart, *Catastrophe Theory and its Applications*. Pitman, London, 1978.

[31] J.R. Rice, On the structure of stress–strain relations for time dependent plastic deformation in metals. *ASME Journal of Applied Mechanics*, **37** (1970), pp. 728–37.

[32] P. Rosakis and J.K. Knowles, Unstable kinetic relations and the dynamics of solid–solid phase transitions. *Journal of the Mechanics and Physics of Solids*, **45** (1997), pp. 2055–81.

[33] M. Shearer, The Riemann problem for a class of conservation laws of mixed type. *Journal of Differential Equations*, **46** (1982), pp. 426–43.

[34] M. Slemrod, Admissibility criteria for propagating phase boundaries in a van der Waals fluid. *Archive for Rational Mechanics and Analysis*, **81** (1983), pp. 301–15.

[35] L.R.G. Treloar, *The Physics of Rubber Elasticity*, third edition. Oxford University Press, Oxford, 1975.

[36] L. Truskinovsky, Structure of an isothermal phase discontinuity. *Soviet Physics Doklady* **30** (1985), pp. 945–8.

[37] N.A. Winfree, *Impact-Induced Phase Changes in Elastic Solids: A Continuum Study Including Numerical Simulations for GeO_2*. Ph.D. dissertation, California Institute of Technology, Pasadena, CA, 1999.

[38] X. Zhong, Dynamic behavior of the interface between two solid phases in an elastic bar. *Journal of Elasticity*, **41** (1995), pp. 39–72.

[39] X. Zhong, T.Y. Hou, and P.G. LeFloch, Computational methods for propagating phase boundaries. *Journal of Computational Physics*, **124** (1996), pp. 192–216.

Part III Thermomechanical Theory

5 Multiple-Well Free Energy Potentials

5.1 Introduction

This, and the next two chapters, address the basic concepts and ingredients underlying the subject of this monograph. Most of these ideas have already been introduced within a one-dimensional setting in the preceding chapters. We now generalize that earlier discussion in two ways: one, we turn to a three-dimensional setting and two, we consider both mechanical and thermal effects.

The present chapter will be concerned with the energy potential. In the one-dimensional purely mechanical setting, we discussed the local minima, or energy wells, of the strain energy $W(\gamma)$ as a function of strain and noted that they corresponded to stress-free configurations of the material. We also discussed the potential energy, $P(\gamma, \sigma) = W(\gamma) - \sigma\gamma$, as a function of strain and stress, and noted that the local minima of $P(\cdot, \sigma)$ corresponded to stress–strain pairs related by the constitutive relation $\sigma = \hat{\sigma}(\gamma) = W'(\gamma)$. In this chapter we generalize both of these notions.

A thermoelastic material can be characterized by a suitable free energy potential. Most of our attention in this chapter will be focused on the local minima – or "energy wells" – of the potential, with particular interest on free energy functions with *multiple* local minima. Such multiwell free energy functions can be used to describe materials that can exist in more than one configuration at certain stress–temperature pairs. We say that each energy well describes a distinct phase of the material, where the term "phase" is to be interpreted broadly. For example, the different energy wells might represent the liquid and vapor phases of a fluid; the austenite and martensite phases of a crystalline solid; the randomly oriented and aligned states of a polymer; or the intact and comminuted states of glass. The phenomenological nature of the theory allows this latitude.

In Section 5.2 we introduce the Helmholtz free energy potential for a thermoelastic continuum and describe its local minima. In order to account for the effect of stress, in Section 5.3 we introduce the notion of the potential energy function and discuss its energy wells and relationship to the Gibbs free energy. These ideas are illustrated in Section 5.4 and Section 5.5 through the respective examples of a van der Waals fluid and a two-phase crystalline solid.

We postpone discussing thermoelasticity more broadly until Chapter 7, our discussion here being limited to the characteristics of the two aforementioned potentials.

5.2 Helmholtz free energy potential

A deformation of a body is described by a function $\mathbf{y}(\mathbf{x})$, which maps the particle located at $\mathbf{x}$ in some reference configuration to a location $\mathbf{y}(\mathbf{x})$ in the deformed configuration. The deformation gradient tensor $\mathbf{F} = \nabla \mathbf{y}$ has positive determinant and can be decomposed uniquely into the product of a proper orthogonal tensor $\mathbf{R}$ and a symmetric positive-definite tensor $\mathbf{U}$ representing, respectively, a rotation and a stretch:

$$\mathbf{F} = \mathbf{RU}. \tag{5.1}$$

The stretch tensor $\mathbf{U}$ is the symmetric positive-definite square root of the right Cauchy–Green tensor[1] $\mathbf{C}$:

$$\mathbf{C} = \mathbf{U}^2 = \mathbf{F}^T\mathbf{F}. \tag{5.2}$$

Various equivalent measures of Lagrangian strain, such as the Green strain,

$$\mathbf{E} = \tfrac{1}{2}(\mathbf{C} - \mathbf{I}), \tag{5.3}$$

can be defined in terms of $\mathbf{C}$.

A thermoelastic material can be characterized by its Helmholtz free energy potential expressed as a function of the deformation gradient tensor $\mathbf{F}$ and the absolute temperature θ:

$$\psi = \widehat{\psi}(\mathbf{F}, \theta); \tag{5.4}$$

ψ represents energy per unit mass. Moreover, the first Piola–Kirchhoff stress tensor σ and the specific entropy per unit mass η in such a material are related to $\mathbf{F}$ and θ through the constitutive relationships

$$\sigma = \rho\widehat{\psi}_{\mathbf{F}}(\mathbf{F}, \theta), \qquad \eta = -\widehat{\psi}_{\theta}(\mathbf{F}, \theta); \tag{5.5}$$

here $\rho > 0$ is the mass density in the reference configuration and the subscripts attached to $\widehat{\psi}$ indicate the gradient with respect to $\mathbf{F}$ and the partial derivative with respect to θ. The Helmholtz free energy ψ is related to the more familiar internal energy (per unit mass) ε by

$$\psi = \varepsilon - \eta\theta; \tag{5.6}$$

---

[1] For a more detailed discussion of the kinematics of a continuum, see, for example, Chadwick [3].

it is worth noting that the relationship (5.6) holds even for materials that are not thermoelastic. For further details on thermoelastic materials, see Section 96 of Truesdell and Noll [10].

Frame indifference requires that the value of the energy remain unchanged by a superposed rigid rotation. This implies that $\widehat{\psi}$ depends on the deformation gradient tensor $\mathbf{F}$ only through the right Cauchy–Green tensor $\mathbf{C}$. Thus we can write

$$\psi = \psi(\mathbf{C}, \theta), \qquad \sigma = 2\rho \, \mathbf{F} \, \psi_{\mathbf{C}}(\mathbf{C}, \theta). \tag{5.7}$$

On the other hand, material symmetry requires that the value of ψ remain unchanged when the reference configuration is subjected to a transformation that preserves the symmetry of the material. Thus $\widehat{\psi}$ must have the further property $\widehat{\psi}(\mathbf{F}, \theta) = \widehat{\psi}(\mathbf{FQ}, \theta)$ for all tensors $\mathbf{F}$ with positive determinant and all tensors $\mathbf{Q}$ in the material's *symmetry group* $\mathcal{G}$. The group $\mathcal{G}$ characterizes the symmetry of the material in the reference configuration. For example, if the material has cubic symmetry in the reference configuration then $\mathcal{G}$ contains the 24 rotations that map a cube back into itself, whereas if it is isotropic, $\mathcal{G}$ contains all rotations. When combined with the implication $(5.7)_1$ of frame indifference, material symmetry requires that, for each symmetric tensor $\mathbf{C}$,

$$\psi(\mathbf{C}, \theta) = \psi(\mathbf{Q}^T \mathbf{C} \mathbf{Q}, \theta) \quad \text{for all} \quad \mathbf{Q} \in \mathcal{G}. \tag{5.8}$$

Consider a configuration of the body, obtained from the reference configuration by a homogeneous deformation with gradient $\mathbf{F}_o$ and corresponding right Cauchy–Green tensor $\mathbf{C}_o$, and suppose that this configuration is *stress free* at the temperature θ_o. Then by $(5.7)_2$ we have $\psi_{\mathbf{C}}(\mathbf{C}_o, \theta_o) = \mathbf{0}$ and so the function $\psi(\cdot, \theta_o)$ has an extremum at $\mathbf{C}_o$. Suppose that this extremum is in fact a local minimum, that is, suppose that

$$\psi(\mathbf{C}', \theta_o) > \psi(\mathbf{C}_o, \theta_o) \quad \text{for all } \mathbf{C}' \text{ near } \mathbf{C}_o, \quad \mathbf{C}' \neq \mathbf{C}_o. \tag{5.9}$$

Then one speaks of $\psi(\cdot, \theta_o)$ as having an *energy well*, and one refers to the tensor $\mathbf{C}_o$ as being at the bottom of this energy well. A homogeneous configuration of the body with associated right Cauchy–Green tensor $\mathbf{C}_o$ is *metastable* in the sense that it corresponds to a local minimum of the free energy; it is *stable* if $\mathbf{C}_o$ is a global minimizer of ψ.

Our interest is in materials that, at some temperature θ_o, possess at least *two* distinct, stress-free metastable equilibrium configurations that do not differ by a rotation. If $\mathbf{C}^+$ and $\mathbf{C}^-$ denote the right Cauchy–Green tensors associated with these two configurations, then $\psi(\cdot, \theta_o)$ must have energy wells at $\mathbf{C}^+$ and $\mathbf{C}^-$. By continuity, the function $\psi(\cdot, \theta)$ will continue to have two energy wells for temperatures close to θ_o. Suppose that the two energy wells have the same height at some temperature θ_T: $\psi(\mathbf{C}^+, \theta_T) = \psi(\mathbf{C}^-, \theta_T)$; θ_T is referred to as the *transformation temperature*. Note that both phases are stable at $\theta = \theta_T$. If $\psi(\mathbf{C}^+, \theta) < \psi(\mathbf{C}^-, \theta)$ for $\theta > \theta_T$, then the plus-phase is stable at temperatures above the transformation temperature and vice versa.

5.3 Potential energy function and the effect of stress

Local minima of the Helmholtz free energy function correspond to stress-free states. As observed previously in the one-dimensional mechanical setting, *stressed* states are local minima of the potential energy function. In three-dimensional thermoelasticity, the potential energy function is defined by

$$\widehat{P}(\mathbf{F}, \theta; \sigma) = \rho \widehat{\psi}(\mathbf{F}, \theta) - \sigma \cdot \mathbf{F} \tag{5.10}$$

where σ is the first Piola–Kirchhoff stress tensor. Note that in the special case of vanishing stress, P coincides with $\rho\psi$.

Frame indifference tells us that there is a function P such that

$$P(\mathbf{C}, \theta; \Sigma) = \widehat{P}(\mathbf{F}, \theta; \sigma) \quad \text{where } \mathbf{C} = \mathbf{F}^T \mathbf{F} \quad \text{and} \quad \Sigma = \mathbf{F}^{-1}\sigma; \tag{5.11}$$

here $\mathbf{C}$ is the right Cauchy–Green tensor encountered previously and the symmetric tensor Σ is the *second Piola–Kirchhoff stress tensor*. Clearly,

$$P(\mathbf{C}, \theta; \Sigma) = \rho\psi(\mathbf{C}, \theta) - \Sigma \cdot \mathbf{C}. \tag{5.12}$$

It is readily seen that the second Piola–Kirchhoff stress Σ obeys the constitutive relation $\Sigma = 2\rho\psi_{\mathbf{C}}(\mathbf{C}, \theta)$.

Given the stress and temperature, suppose that the associated potential energy function $P(\cdot, \theta; \Sigma)$ has an extremum at some $\mathbf{C}$. Then one sees from (5.12) that the triplet $\{\mathbf{C}, \theta, \Sigma\}$ satisfies the constitutive law $\Sigma = 2\rho\psi_{\mathbf{C}}(\mathbf{C}, \theta)$. Extrema of $P(\cdot, \theta; \Sigma)$ therefore describe states that obey the thermoelastic constitutive equation for stress. If this extremum is in fact a local minimum, then in addition $\psi_{\mathbf{CC}}(\mathbf{C}, \theta)$ must be positive definite. A state $\{\mathbf{C}, \theta, \Sigma\}$ is said to be *metastable* if $\mathbf{C}$ is a local minimizer of the potential energy $P(\cdot, \theta; \Sigma)$ and *stable* if it is the global minimizer; in either case, we say that P has an energy well and that $\mathbf{C}$ is located at the bottom of this energy well. Note that these notions of metastability, stability, and energy wells are consistent with our previous discussion pertaining to the special case of vanishing stress.

Suppose that at some particular (θ, Σ) the potential energy function $P(\cdot, \theta; \Sigma)$ has two energy wells, one at $\mathbf{C} = \mathbf{C}^+$ and the other at $\mathbf{C} = \mathbf{C}^-$. These minima can be associated with two phases of the material, both of which exist at the given stress and temperature. The height between these energy wells is

$$h = P(\mathbf{C}^+, \theta; \Sigma) - P(\mathbf{C}^-, \theta; \Sigma); \tag{5.13}$$

if $h > 0$ the phase characterized by $\{\mathbf{C}^-, \theta, \Sigma\}$ is the global minimizer of the potential energy and so is stable in the preceding sense, whereas if $h < 0$ then $\{\mathbf{C}^+, \theta, \Sigma\}$ is stable. Note from (5.13), (5.11)$_1$, and (5.10) that we can write

$$h = \rho\widehat{\psi}(\mathbf{F}^+, \theta) - \rho\widehat{\psi}(\mathbf{F}^-, \theta) - \sigma^+ \cdot \mathbf{F}^+ + \sigma^- \cdot \mathbf{F}^-. \tag{5.14}$$

In certain special cases the height h is related to the driving force f to be encountered in Chapter 7.

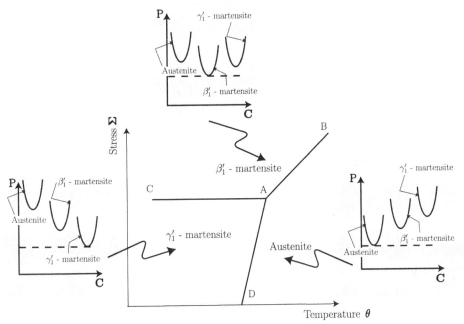

Figure 5.1. Schematic phase diagram for the three phases of a CuAlNi shape memory alloy. The inset figures show sketches of the potential energy function at different points (θ, Σ) on the phase diagram.

It should be noted that even if the Helmholtz function itself has only a single energy well, it is still possible for the potential energy function P to have multiple energy wells. A material characterized by such a ψ exhibits a single phase when it is unstressed, but exhibits multiple phases when suitably stressed. Whether or not this is so depends on the convexity of ψ.

The number and character of the local minima of the potential energy $P(\cdot, \theta; \Sigma)$ depends on (θ, Σ) and it is often useful to characterize the minima in the temperature, stress "plane"; when used for this purpose, the (θ, Σ) plane provides what is called the *phase diagram* of the material. Consider, for example, an alloy of CuAlNi at compositions where it is a shape memory material. This alloy involves an austenite phase that is cubic, one martensite phase that is orthorhombic (called γ_1'), and a second martensite phase that is monoclinic (called β_1'); in subsequent chapters we shall have occasion to encounter all three of these phases. Depending on the stress and temperature, a different one of these phases is energetically favorable. Figure 5.1 shows a schematic phase diagram for this material, indicating, at each point (θ, Σ), what the lowest-energy phase is. For example, γ_1'-martensite is seen to be energetically preferred at points below AC and to the left of AD corresponding to (relatively) low stress and low temperature and so on.

The potential energy function $P(\mathbf{C}, \theta; \Sigma)$ is related to, though not identical with, the Gibbs free energy function g. Since g is expressed as a function of stress and temperature alone, in order to explore the relationship between P and g it is necessary to eliminate the right Cauchy–Green tensor in P in favor of stress and temperature. If the constitutive relation $\boldsymbol{\Sigma} = 2\rho\psi_{\mathbf{C}}(\mathbf{C}, \theta)$ for the second Piola–Kirchhoff stress tensor $\boldsymbol{\Sigma}$ is invertible, then we can write $\mathbf{C} = \tilde{\mathbf{C}}(\boldsymbol{\Sigma}, \theta)$. The potential g defined by

$$g(\boldsymbol{\Sigma}, \theta) = P(\tilde{\mathbf{C}}(\boldsymbol{\Sigma}, \theta), \theta; \boldsymbol{\Sigma}) = \rho\psi(\tilde{\mathbf{C}}(\boldsymbol{\Sigma}, \theta), \theta) - \boldsymbol{\Sigma} \cdot \tilde{\mathbf{C}}(\boldsymbol{\Sigma}, \theta) \quad (5.15)$$

represents the *Gibbs free energy* per unit reference volume. Observe that the *value* of the potential energy *at an extremum* of $P(\cdot, \theta; \boldsymbol{\Sigma})$ coincides with the corresponding value of the Gibbs free energy at that same $(\theta, \boldsymbol{\Sigma})$.

We now turn to two examples that illustrate the concepts outlined here and in the previous section.

5.4 Example 1: the van der Waals fluid

One of the simplest examples of a thermoelastic material is an inviscid, compressible fluid (i.e. a "thermoelastic fluid"). Such a material is characterized by a Helmholtz free energy potential of the special form:

$$\psi(\mathbf{C}, \theta) = \psi(v, \theta), \quad \text{where} \quad v = v_o J = v_o\sqrt{\det \mathbf{C}} \quad (5.16)$$

is the specific volume and $v_o = 1/\rho$ is the specific volume in the reference configuration. It follows from (5.16), (5.5)$_1$, and $\partial v/\partial \mathbf{F} = v_o J(\mathbf{F}^{-1})^T$ that the true stress, $\boldsymbol{\tau} = (\det \mathbf{F})^{-1}\boldsymbol{\sigma}\mathbf{F}^T$, in such a fluid is necessarily hydrostatic, that is, $\boldsymbol{\tau} = -p\mathbf{I}$, where the pressure p is given by

$$p = -\frac{\partial \psi}{\partial v}. \quad (5.17)$$

The classical example of a thermoelastic fluid is an *ideal gas*. This is characterized by the Helmholtz free energy function

$$\psi(v, \theta) = -c_v\theta \, \log\left(\frac{\theta}{\theta_o}\right) - R\theta \, \log v, \quad v > 0, \quad \theta > 0, \quad (5.18)$$

where c_v, θ_o, and R are positive material constants. Equations (5.17) and (5.18) lead to the well-known ideal gas law

$$p = \frac{R\theta}{v}, \quad (5.19)$$

and from $\epsilon = \psi - \theta\partial\psi/\partial\theta$ and (5.18) one recovers the familiar expression

$$\epsilon = c_v\theta \quad (5.20)$$

for the internal energy per unit mass of an ideal gas.

The ideal gas law $p = R\theta/v$ can be derived from molecular-scale arguments based on the kinetic theory of gases. Among the various assumptions underlying this derivation is one that treats the gas as a collection of point masses and another that neglects any forces between pairs of molecules. An intuitive way in which to correct for the former approximation is to replace v in the ideal gas law by $v - b$ where the empirical constant $b > 0$ accounts for the volume occupied by the gas molecules themselves. With regard to the second assumption, the presence of inter-molecular forces will cause an effective backwards force on a molecule as it approaches a wall of the container, and therefore, its effect would be to reduce the pressure in proportion to the number of interacting pairs of molecules. This suggests that the value of p given by the ideal gas law should be decreased by a term proportional to $1/v^2$.

These heuristic arguments suggest that a better description of a fluid is provided by the *van der Waals model*

$$p = p(v, \theta) = \frac{R\theta}{v - b} - \frac{a}{v^2}, \quad v > b, \quad \theta > 0. \tag{5.21}$$

A more detailed discussion of the ideas leading to (5.21) can be found, for example, in Section 20.4 of Kestin [6]. A complete characterization of a van der Waals fluid is provided by the Helmholtz free energy function

$$\psi(v, \theta) = -c_v\theta \log\left(\frac{\theta}{\theta_o}\right) - R\theta \log(v - b) - \frac{a}{v}, \quad v > b, \quad \theta > 0, \tag{5.22}$$

where c_v, a, b, R, θ_o, and v_o are positive constants and $v = v_o J = v_o\sqrt{\det \mathbf{C}}$ is the specific volume; see equations (3.49) and (3.50) of Callen [2] from which (5.22) can be derived. Expressions for other physical quantities can be derived from (5.22). For example (5.17) can be used to calculate the pressure, leading back to (5.21), while the specific internal energy can be determined using $\epsilon = \psi - \theta\partial\psi/\partial\theta$, which gives

$$\epsilon = c_v\theta - \frac{a}{v}. \tag{5.23}$$

Numerical values of the parameters a, b, and c_v for which the van der Waals model reasonably describes a variety of fluids have been reported in the literature; see for example Table 3.1 of Callen [2].

In order to examine the ability of the van der Waals model to model multiple phases, we consider the potential energy function

$$P(v, \theta; p) = \psi(v, \theta) + pv = -c_v\theta \log\left(\frac{\theta}{\theta_o}\right) - R\theta \log(v - b) - \frac{a}{v} + pv, \tag{5.24}$$

which is defined for all $v > b$, $\theta > 0$, and $-\infty < p < \infty$. For certain values of temperature θ and pressure p one can readily show that the function $P(\cdot, \theta; p)$ has two local minima corresponding to two phases of the fluid. For other values of θ and p it has a single minimum corresponding to one phase. Analysis of (5.24) shows that the detailed character of $P(\cdot, \theta; p)$ is as depicted in Figure 5.2: P has two local minima when the pressure and temperature correspond to a point in the shaded region enclosed by the curves $p = p_M(\theta)$, $p = p_m(\theta)$, $p = 0$, and $\theta = 0$. Explicit expressions for the functions $p_m(\theta)$ and $p_M(\theta)$ can be determined but

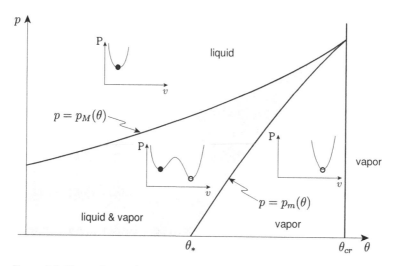

Figure 5.2. Phase diagram for a van der Waals fluid on the temperature, pressure plane. The line $\theta = \theta_{cr}$ and the curves $p = p_M(\theta)$ and $p = p_m(\theta)$ divide this plane into three distinct regions. The number of local minima of the potential energy function $P(\cdot, \theta; p)$ at points corresponding to each of these subregions is shown in the inset figures. The local minima of P correspond to the liquid (filled circles) and vapor (open circles) phases of the fluid.

are not displayed; the two temperatures θ_{cr} and θ_* shown in the figure have the values

$$\theta_{cr} = \frac{8}{27} \frac{a}{bR}, \qquad \theta_* = \frac{27}{32} \theta_{cr}. \tag{5.25}$$

Of the two local minima, the one at the smaller value of v corresponds to the more dense phase, and it is natural therefore to refer to it as the liquid phase; the other minimum describes the vapor phase. When the temperature and pressure are associated with either $\{(\theta, p) : 0 < \theta < \theta_{cr}, \ p > p_M(\theta)\}$ or $\{(\theta, p) : \theta_* < \theta < \theta_{cr}, \ p < p_m(\theta)\}$ the potential energy has a single minimum; in the former case this corresponds to the liquid phase, while in the latter it is associated with the vapor phase.

Consider for instance a van der Waals fluid at some fixed temperature in the interval $\theta_* < \theta < \theta_{cr}$. If the pressure is gradually decreased from an initial value that exceeds $p_M(\theta)$, we see from Figure 5.2 that the fluid is first in its liquid phase and eventually in its vapor phase. This does not tell us the value of pressure at which the fluid changes phase. All we do know in this regard is that on the intermediate interval of pressure, $p_m(\theta) < p < p_M(\theta)$, both phases are possible.

Next we examine the stability of each phase associated with a point in the two-phase portion of the θ, p-plane. To do this, we need to determine which of the two local minima of the potential energy P is the global minimizer. Analysis of (5.24) shows that there is a pressure level $p_0(\theta)$, intermediate to $p_M(\theta)$ and $p_m(\theta)$, such that the local minimum at the smaller value of v, that is, the local minimum associated with the liquid phase, is the global minimum for $p > p_0(\theta)$. The vapor phase is stable for $p < p_0(\theta)$. The pressure $p_0(\theta)$, at which both phases

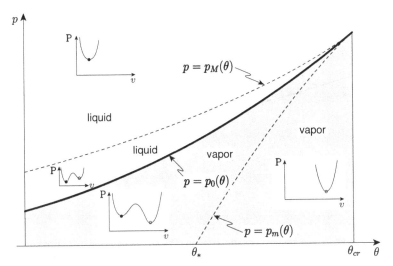

Figure 5.3. The stable phases of a van der Waals fluid on the temperature, pressure plane. The filled and open circles correspond respectively to the liquid and vapor phases.

are stable, is known as the *Maxwell pressure*. This is depicted in Figure 5.3 in which the vapor phase is stable on the shaded region and the liquid phase is stable on the unshaded region. Again, suppose that the temperature is fixed in the interval $\theta_* < \theta < \theta_{cr}$ and that the pressure is gradually decreased from an initial value, which exceeds $p_M(\theta)$. Suppose further that the fluid assumes the liquid or vapor phase according to which of these phases is the stable one. Then, the fluid begins in the liquid phase and transforms to the vapor phase when the pressure falls below the value $p_0(\theta)$.

Finally it is worth describing these observations with reference to the isothermal pressure–volume curve. In contrast to an ideal gas, the p, v-curve for a van der Waals fluid does not always decline monotonically. If the temperature lies in the range $\theta > \theta_{cr}$, one finds that p is in fact a monotonically decreasing function of v just as for an ideal gas; see Figure 5.4. However for temperatures in the range $0 < \theta < \theta_{cr}$, one finds that p decreases with increasing v on two intervals, $b < v < v_m(\theta)$ and $v > v_M(\theta)$, but increases on the intermediate interval $v_m(\theta) < v < v_M(\theta)$. Here $v_m(\theta)$ and $v_M(\theta)$ are given by the two roots $(> b)$ of the cubic equation

$$\left(\frac{4}{27} \frac{\theta}{\theta_{cr}} \right) v^3 = b(v - b)^2. \tag{5.26}$$

The corresponding values of pressure, denoted by $p_m(\theta)$ and $p_M(\theta)$ in Figure 5.4, are given by $p_m(\theta) = p(v_m(\theta), \theta)$ and $p_M(\theta) = p(v_M(\theta), \theta)$ where $p(v, \theta)$ is given by (5.21). Graphs of the functions $p_M(\theta)$ and $p_m(\theta)$ were shown previously in Figure 5.2.

Next, observe from (5.24) that if $P(\cdot, \theta; p)$ has an extremum at some v, then the triplet $\{v, \theta, p\}$ satisfies the pressure–volume–temperature constitutive relation $p = p(v, \theta) = -\partial \psi(v, \theta)/\partial v$. Moreover, since $\partial^2 P(v, \theta; p)/\partial v^2 = -\partial p(v, \theta)/\partial v$, this

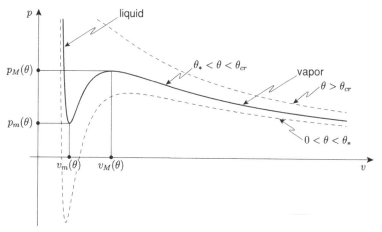

Figure 5.4. Graphs of pressure versus specific volume at three different fixed values of temperature for a van der Waals fluid. The declining branches correspond to the liquid and vapor phases of the fluid.

extremum is a local minimum if and only if $\partial p(v, \theta)/\partial v < 0$. Consequently each declining branch of the isothermal p, v-curve can be associated with a local minimum of P and therefore corresponds to a metastable state of the fluid; the rising branch is unstable. For the van der Waals fluid, the two declining branches can therefore be identified with the liquid and vapor phases, the branch to the left (associated with the smaller values of v) being associated with the liquid phase.

Finally, the Maxwell pressure $p_0(\theta)$ can be shown to cut off equal areas of the p, v-curve as shown in Figure 5.5. Since we know that the liquid phase is stable for $p > p_0(\theta)$ and the vapor phase is stable for $p < p_0(\theta)$, the stable phases correspond to the bold portions of the p, v-curve in Figure 5.5; the metastable phases are associated with the light portion.

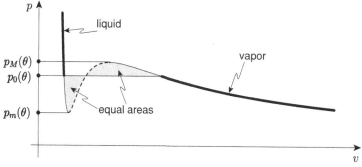

Figure 5.5. Graph of the pressure p versus specific volume v for a van der Waals fluid at a fixed temperature in the range $0 < \theta < \theta_{cr}$. The bold portions of the curve correspond to the stable phase while the light portion is associated with the metastable phase. The dashed portion of the curve is unstable. The liquid phase is stable above the Maxwell pressure $p_0(\theta)$ and metastable below it; the vapor phase is stable below the Maxwell pressure and metastable above it. The Maxwell pressure $p_0(\theta)$ has the property that the two shaded regions in the figure have equal area.

5.5 Example 2: two-phase martensitic material with cubic and tetragonal phases

Example 2 concerns a crystalline solid. We study the kinematics of the transformation by treating the crystal as a lattice of atoms and then use the results from that analysis in developing a continuum model. The model developed here is due to Ericksen and the reader can find a more detailed discussion in [4].

Consider a two-phase crystalline solid in the absence of stress. The associated lattice has one structure at high temperatures and a different one at low temperatures, the change between them being sudden (a "first-order martensitic phase transformation"). The high-temperature lattice is characterized by one set of lattice vectors, say $\{e_1^a, e_2^a, e_3^a\}$, and describes one phase of the material usually called *austenite*; the low-temperature lattice is described by $\{e_1^m, e_2^m, e_3^m\}$ and corresponds to a second phase of the material, *martensite*. Since the two lattices are different, $\{e_1^a, e_2^a, e_3^a\} \neq \{e_1^m, e_2^m, e_3^m\}$. Thus as the temperature of such a material is changed, the lattice vectors first vary slowly and continuously due to thermal expansion, and then suddenly, at the *transformation temperature* θ_T, they change *discontinuously*.

Since the change in the lattice due to thermal expansion is much smaller than that due to phase transformation (at least in the range of temperatures we shall be concerned with), we will neglect thermal expansion. Both sets of lattice vectors are therefore assumed to be independent of temperature in this chapter.

Examples of such materials include the alloys gold–cadmium (cubic and orthorhombic phases), nickel–titanium (cubic and monoclinic phases), and copper–aluminum–nickel (cubic, orthorhombic, and monoclinic phases). One of the simplest examples of a two-phase martensitic material is one that occurs in face-centered cubic and face-centered tetragonal phases, For example, indium–thallium, manganese–nickel, and manganese–copper. Figure 5.6 depicts the lattices associated with such a material; for more details, see, for example Bhattacharya [1].

Let $\{c_1, c_2, c_3\}$ denote fixed *unit* vectors in the cubic directions and let the lattice vectors associated with the austenite lattice be denoted by $\{e_1^a, e_2^a, e_3^a\}$. If $a_o \times a_o \times a_o$ denotes the lattice parameters of the cubic phase then the lattice vectors shown in Figure 5.6(a) are

$$e_1^a = \frac{a_o}{2}(c_2 + c_3), \qquad e_2^a = \frac{a_o}{2}(-c_2 + c_3), \qquad e_3^a = \frac{a_o}{2}(c_1 + c_3). \quad (5.27)$$

Similarly for the tetragonal phase, if the lattice parameters are denoted by $a \times a \times c$, the lattice vectors $\{e_1^m, e_2^m, e_3^m\}$ can be taken to be

$$e_1^m = \frac{a}{2}(c_2 + c_3), \qquad e_2^m = \frac{a}{2}(-c_2 + c_3), \qquad e_3^m = \frac{c}{2}c_1 + \frac{a}{2}c_3. \quad (5.28)$$

The two sets of lattice vectors $\{e_1^a, e_2^a, e_3^a\}$ and $\{e_1^m, e_2^m, e_3^m\}$ are each linearly independent. Therefore there is a nonsingular linear transformation U_1 that relates them:

$$e_i^m = U_1 e_i^a . \quad (5.29)$$

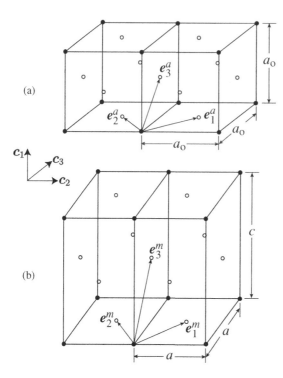

Figure 5.6. (a) Face-centered cubic lattice with lattice parameters $a_o \times a_o \times a_o$ and lattice vectors $\{\mathbf{e}_1^a, \mathbf{e}_2^a, \mathbf{e}_3^a\}$, and (b) face-centered tetragonal lattice with lattice parameters $a \times a \times c$ and lattice vectors $\{\mathbf{e}_1^m, \mathbf{e}_2^m, \mathbf{e}_3^m\}$. The unit vectors $\{\mathbf{c}_1, \mathbf{c}_2, \mathbf{c}_3\}$ are associated with the cubic directions. Solely for clarity, the atoms at the vertices are depicted by (small) filled circles while those at the centers of the faces are shown by open circles.

Consequently we may regard the martensite lattice as the image of the austenite lattice under the linear transformation $\mathbf{U}_1$. A characteristic of martensitic transformations is that the change of crystal structure is diffusionless, that is, there is no rearrangement of atoms associated with the lattice change. Consequently, we can describe the kinematics of the phase transformation completely by the linear transformation $\mathbf{U}_1$. The tensor $\mathbf{U}_1$ is called the *Bain stretch* or *transformation stretch*; observe that knowledge of the austenitic and martensitic lattices determines $\mathbf{U}_1$.

One can readily verify from (5.27) and (5.28) that in the present example,

$$\mathbf{U}_1 = \eta_3 \, \mathbf{c}_1 \otimes \mathbf{c}_1 + \eta_1 \, \mathbf{c}_2 \otimes \mathbf{c}_2 + \eta_1 \, \mathbf{c}_3 \otimes \mathbf{c}_3 \qquad (5.30)$$

where the stretches η_1 and η_3 are given by

$$\eta_1 = \frac{a}{a_o}, \qquad \eta_3 = \frac{c}{a_o}. \qquad (5.31)$$

Thus the transformation from the cubic lattice to the tetragonal lattice is achieved by stretching the parent lattice equally in two of the cubic directions and unequally in the third direction, the stretches associated with this being η_1, η_1, and η_3.

We now construct a continuum model of this crystalline material. Suppose that a stress-free configuration of the cubic phase (austenite) is taken as the reference configuration of the continuum; the associated lattice can be characterized by the lattice vectors $\{\mathbf{e}_1^a, \mathbf{e}_2^a, \mathbf{e}_3^a\}$. Next consider an arbitrary second configuration of the continuum, and let $\{\mathbf{e}_1, \mathbf{e}_2, \mathbf{e}_3\}$ denote a set of lattice vectors characterizing the lattice associated with this second configuration. Let $\mathbf{F}$ denote the unique nonsingular tensor that relates these two sets of lattice vectors:

$$\mathbf{e}_k = \mathbf{F}\mathbf{e}_k^a, \quad k = 1, 2, 3. \tag{5.32}$$

Then, according to the Cauchy–Born rule, the gradient of the deformation relating the two continuum configurations is also $\mathbf{F}$. Therefore in particular, if the reference configuration is subjected to a homogeneous deformation with deformation gradient tensor $\mathbf{U}_1$, we obtain, according to the preceding discussion, a second stress-free configuration that corresponds to the tetragonal phase (martensite). The deformation gradient tensors $\mathbf{F} = \mathbf{I}$ and $\mathbf{F} = \mathbf{U}_1$ therefore each describes a stress-free equilibrium configuration.

In the model we develop here, we express the Helmholtz free energy ψ as a function of the Green strain $\mathbf{E}$: $\psi = \psi(\mathbf{E}, \theta)$ where $\mathbf{E} = (1/2)(\mathbf{F}^T\mathbf{F} - \mathbf{I})$. Since both phases exist at the transformation temperature θ_T, $\psi(\cdot, \theta_T)$ must have local minima at $\mathbf{E} = \mathbf{0}$ and $\mathbf{E} = \mathbf{E}_1 = (1/2)(\mathbf{U}_1^2 - \mathbf{I})$. By continuity, $\psi(\cdot, \theta)$ will continue to have local minima at these same values[2] of $\mathbf{E}$ for temperatures close to the transformation temperature.

Since the austenite phase is preferred over the martensite phase for temperatures greater than θ_T, this implies that the cubic phase is the low-energy phase for $\theta > \theta_T$, that is, that $\psi(\mathbf{0}, \theta) < \psi(\mathbf{E}_1, \theta)$ for $\theta > \theta_T$. Similarly, since the martensite phase is favored for temperatures less than θ_T we must also have $\psi(\mathbf{E}_1, \theta) < \psi(\mathbf{0}, \theta)$ for $\theta > \theta_T$. By continuity, both phases must have equal energy at the transformation temperature: $\psi(\mathbf{E}_1, \theta_T) = \psi(\mathbf{0}, \theta_T)$.

We now construct an explicit free energy function $\psi(\mathbf{E}, \theta)$, which possesses the properties outlined above. Since the reference configuration corresponds to an unstressed state of austenite, the free energy function $\psi(\mathbf{E}, \theta)$ must possess cubic symmetry. It is known (see, e.g., Smith and Rivlin [8], and Green and Adkins [5]) that, to have cubic symmetry, ψ depends on $\mathbf{E}$ only through the "cubic invariants"

$$
\begin{aligned}
I_1 &= E_{11} + E_{22} + E_{33}, \\
I_2 &= E_{11}E_{22} + E_{22}E_{33} + E_{33}E_{11}, \\
I_3 &= E_{11}E_{22}E_{33}, \\
I_4 &= E_{12}E_{23}E_{31}, \\
I_5 &= E_{12}^2 + E_{23}^2 + E_{31}^2, \\
I_6 &= (E_{11} + E_{22})E_{12}^2 + (E_{22} + E_{33})E_{23}^2 + (E_{33} + E_{11})E_{31}^2, \\
I_7 &= E_{12}^2E_{23}^2 + E_{23}^2E_{31}^2 + E_{31}^2E_{12}^2, \\
I_8 &= E_{11}E_{12}^2E_{23}^2 + E_{22}E_{23}^2E_{31}^2 + E_{33}E_{31}^2E_{12}^2, \\
I_9 &= E_{11}E_{22}E_{12}^2 + E_{22}E_{33}E_{23}^2 + E_{33}E_{11}E_{31}^2,
\end{aligned}
\tag{5.33}
$$

---

[2] We are neglecting the effect of thermal expansion here.

where E_{ij} refers to the i, j-component of $\mathbf{E}$ in the cubic basis $\{\mathbf{c}_1, \mathbf{c}_2, \mathbf{c}_3\}$. The number of invariants in this list depends on the particular cubic class under consideration (e.g. Section 1.11 of Green and Adkins [5]) but the analysis below is valid for all of these classes.

For temperatures close to the transformation temperature, Ericksen [4] has argued on the basis of experimental observations that as a first approximation, all of the shear strain components in the cubic basis vanish, and additionally, that the sum of the normal strains also vanishes. Accordingly he suggested a geometrically constrained theory based on the two constraints

$$
\begin{aligned}
I_1 &= E_{11} + E_{22} + E_{33} = 0, \\
I_5 &= E_{12}^2 + E_{23}^2 + E_{31}^2 = 0.
\end{aligned}
\tag{5.34}
$$

In this case, the only two nontrivial strain invariants among those in the preceding list are I_2 and I_3 and so $\psi = \psi(I_2, I_3, \theta)$. Suppose that ψ is a polynomial in the components of $\mathbf{E}$. In order to capture the desired multiwell character, ψ must be at least a quartic polynomial. It follows from (5.33) and (5.34) that the most general quartic polynomial of the form $\psi = \psi(I_2, I_3, \theta)$ is

$$
\psi = c_0 + c_2 I_2 + c_3 I_3 + c_{22} I_2^2,
\tag{5.35}
$$

where the coefficients c_i may depend on temperature: $c_i = c_i(\theta)$.

We now impose the requirement that $\psi(\cdot, \theta_T)$ must have local minima at $\mathbf{E} = \mathbf{0}$ and $\mathbf{E} = \mathbf{E}_1 = (1/2)(\mathbf{U}_1^2 - \mathbf{I})$ where $\mathbf{U}_1$ is given by (5.30). The function ψ given in (5.35) automatically has an extremum at $\mathbf{E} = \mathbf{0}$, and this is a local minimum if $c_2 < 0$. In order for ψ to have an extremum at $\mathbf{E} = \mathbf{E}_1$ it is necessary that $c_2 = -pc_3 + 6p^2 c_{22}$, where we have set[3]

$$
p = \frac{1}{2}(\eta_1^2 - 1).
\tag{5.36}
$$

If this extremum is to be a local minimum one must also have $12pc_{22} > c_3 > 0$ where we are considering the case $p > 0$. Combining these various requirements leads to

$$
c_2 = -pc_3 + 6p^2 c_{22}, \qquad 12pc_{22} > c_3 > 6pc_{22} > 0.
\tag{5.37}
$$

Finally we must ensure that the energies of the two stress-free states, $\psi(\mathbf{0}, \theta)$ and $\psi(\mathbf{E}_1, \theta)$, are ordered appropriately as functions of θ. First, note that $\psi(\mathbf{0}, \theta) = c_o(\theta)$ and $\psi(\mathbf{E}_1, \theta) = c_o(\theta) + p^3[c_3(\theta) - 9pc_{22}(\theta)]$. Since the martensite minimum must be lower than the austenite minimum for low temperatures and vice versa for high temperatures, and the values of ψ at the two minima must be the same at the transformation temperature, we require that

$$
c_3'(\theta) - 9pc_{22}'(\theta) > 0, \qquad c_3(\theta_T) = 9pc_{22}(\theta_T),
\tag{5.38}
$$

---

[3] Note that the constraint $I_1 = 0$ implies that $\operatorname{tr} \mathbf{E}_1 = 0$ and therefore that the lattice stretches η_1 and η_3 must be related by $2\eta_1^2 + \eta_3^2 = 3$.

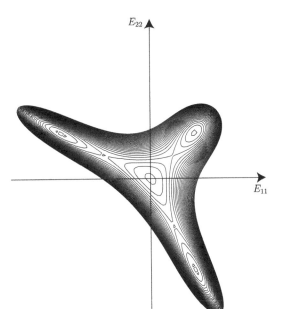

Figure 5.7. Constant energy contours of the free energy (5.39) on the E_{11}, E_{22}-plane at the fixed temperature $\theta = \theta_T$. Observe the presence of the austenitic energy well at $\mathbf{E} = \mathbf{0}$ surrounded by the three martensitic energy wells at $\mathbf{E} = \mathbf{E}_1, \mathbf{E}_2, \mathbf{E}_3$.

where the prime denotes differentiation with respect to the argument and θ_T is the transformation temperature.

In summary, by collecting all of the preceding requirements and introducing a more convenient pair of coefficients $d(\theta)$ and $e(\theta)$, the free energy function ψ can be expressed as

$$\psi = c_o(\theta) + d(\theta)\left[\frac{I_3}{p^3} - \frac{I_2}{p^2}\right] + e(\theta)\left[\left(\frac{I_2}{p^2}\right)^2 - 3\frac{I_2}{p^2} + 9\frac{I_3}{p^3}\right], \tag{5.39}$$

$$e(\theta) > 0, \quad 3e(\theta) > d(\theta) > -3e(\theta), \quad d'(\theta) > 0, \quad d(\theta_T) = 0,$$

where the invariants I_2 and I_3 are given by (5.33). Figure 5.7 shows a contour plot of the free energy (5.39). In view of the constraints (5.34) there are only two independent strain components, which we have taken to be E_{11} and E_{22}. The figure shows the contours of $\psi(E_{11}, E_{22}, \theta)$, as given by (5.39), on the E_{11}, E_{22}-plane. It shows the austenitic energy well at the origin $\mathbf{E} = \mathbf{0}$ surrounded by three martensitic energy wells at $\mathbf{E} = \mathbf{E}_1, \mathbf{E}_2,$ and $\mathbf{E}_3$. The values of the material parameters associated with this figure were chosen solely on the basis of obtaining a fairly clear contour plot.

Although we only required ψ to have two local minima, one at $\mathbf{E} = \mathbf{0}$ and the other at $\mathbf{E} = \mathbf{E}_1$, we see from Figure 5.7 that, in fact, it has four local minima. In order to understand this, observe from (5.8) that necessarily

$$\psi(\mathbf{E}_1, \theta) = \psi(\mathbf{Q}^T \mathbf{E}_1 \mathbf{Q}, \theta) \quad \text{for all } \mathbf{Q} \in \mathcal{G}_a \tag{5.40}$$

where $\mathcal{G}_a$ is the symmetry group of the reference configuration, that is, of unstressed austenite. Since $\mathbf{E}_1$ yields the global minimum of the free energy for $\theta < \theta_T$, it

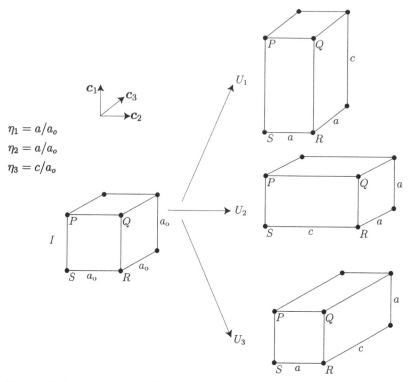

Figure 5.8. Cubic austenite and three variants of tetragonal martensite. Observe that one *cannot* rigidly rotate a martensite variant in such a way as to make the atoms P, Q, R, S coincide with the locations of these same atoms in any other martensite variant.

follows from (5.40) that the tensor $\mathbf{Q}^T\mathbf{E}_1\mathbf{Q}$ is also a minimizer for each $\mathbf{Q} \in \mathcal{G}_a$. As $\mathbf{Q}$ varies over the set of all tensors in $\mathcal{G}_a$, the tensor $\mathbf{Q}^T\mathbf{E}_1\mathbf{Q}$ takes on a certain finite number of distinct, symmetric tensor values, which we denote by $\mathbf{E}_1, \mathbf{E}_2, \ldots, \mathbf{E}_N$, and ψ has local minima at each of these $\mathbf{E}_k$s. These stretch tensors are said to describe the N *variants* of martensite. Note that because of (5.40), necessarily, all variants have the same energy:

$$\psi(\mathbf{E}_1, \theta) = \psi(\mathbf{E}_2, \theta) = \cdots = \psi(\mathbf{E}_N, \theta). \qquad (5.41)$$

The number N depends on the difference in material symmetry between austenite and martensite and in general is larger the greater this difference. When the austenite phase is cubic, the number of martensitic variants is, respectively, 3, 6, or 12 depending on whether the martensite is tetragonal, orthorhombic, or monoclinic. Figure 5.8 illustrates the three variants of tetragonal martensite. Observe that there are three possible ways in which to stretch the cubic lattice in order to obtain the tetragonal lattice, $\mathbf{U}_1$ being the one where the unequal stretching occurs in the $\mathbf{c}_1$ direction. More generally, one can stretch the cubic lattice by the ratio η_3 in the $\mathbf{c}_k$ direction and by η_1 in the remaining two cubic directions; this leads to the three Bain stretch tensors

$$\mathbf{U}_k = \eta_1\mathbf{I} + [\eta_3 - \eta_1]\mathbf{c}_k \otimes \mathbf{c}_k, \quad k = 1, 2, 3, \qquad (5.42)$$

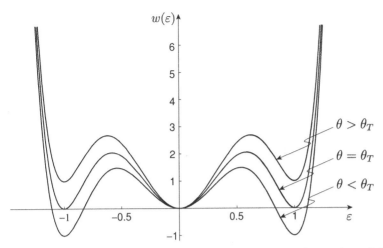

Figure 5.9. A one-dimensional cross-section of the energy (5.39) along the path (5.44) in strain space. The figure plots $w(\zeta) = \psi(\mathbf{E}(\zeta), \theta)$ versus strain ζ at three different temperatures. Because we have omitted the effect of thermal expansion, the locations of the local minima do not change with temperature. The material parameters underlying this plot are the same as those associated with Figure 5.7.

whose matrices of components in the cubic basis are

$$
\underline{U}_1 = \begin{pmatrix} \eta_3 & 0 & 0 \\ 0 & \eta_1 & 0 \\ 0 & 0 & \eta_1 \end{pmatrix}, \qquad
\underline{U}_2 = \begin{pmatrix} \eta_1 & 0 & 0 \\ 0 & \eta_3 & 0 \\ 0 & 0 & \eta_1 \end{pmatrix}, \qquad
\underline{U}_3 = \begin{pmatrix} \eta_1 & 0 & 0 \\ 0 & \eta_1 & 0 \\ 0 & 0 & \eta_3 \end{pmatrix}. \qquad (5.43)
$$

In order to draw a one-dimensional cross-section of the energy (5.39), consider the path in strain space $\mathbf{E} = \mathbf{E}(\zeta)$, $-1.5 < \zeta < 1.5$, where

$$
\underline{E}(\zeta) = \begin{pmatrix} p\zeta^2 & 0 & 0 \\ 0 & p\zeta(3-\zeta)/2 & 0 \\ 0 & 0 & -p\zeta(3+\zeta)/2 \end{pmatrix}. \qquad (5.44)
$$

Observe that $\mathbf{E}(0) = \mathbf{0}$, $\mathbf{E}(-1) = \mathbf{E}_2$, $\mathbf{E}(1) = \mathbf{E}_3$ so that this path passes through the austenite well and two of the martensite wells. Figure 5.9 shows the variation of energy along this path at three different temperatures: the figure displays $w(\zeta) = \psi(\mathbf{E}(\zeta), \theta)$ versus ζ. The three graphs correspond to three temperatures, greater than, equal to, and less than the transformation temperature θ_T. In drawing this figure, we took the coefficient $c_3(\theta)$ in (5.35) to depend linearly on temperature and $c_{22}(\theta)$ to be independent of temperature; the specific values of the various parameters were chosen solely on the basis of obtaining a clear figure.

Due to the restrictive nature of the kinematic constraints (5.34), the free energy function (5.35) cannot provide a *quantitatively* accurate model, especially if the temperature departs from the transformation temperature. The natural generalization of (5.35) is therefore to relax these constraints in the usual way by using

Lagrange multipliers and to replace (5.35) by

$$\psi = d_0 + d_2 \left(I_2 - \tfrac{1}{3}I_1^2\right) + d_3 \left(I_3 - \tfrac{1}{3}I_1 I_2 + \tfrac{2}{27}I_1^3\right) + d_4 \left(I_2 - \tfrac{1}{3}I_1^2\right)^2$$
$$+ d_5 I_5 + d_6 I_1^2 + d_7 I_1^4, \tag{5.45}$$

where $d_i = d_i(\theta)$. Observe that the coefficients of d_2, d_3, and d_4 are quadratic, cubic, and quartic respectively and that they coincide with the respective coefficients of c_2, c_3, and c_4 when the constraints hold. In fact, these coefficients are simply the invariants I_2, I_3, and I_2^2 based on the deviatoric strain tensor $\mathbf{E} - (\mathrm{tr}\mathbf{E}/3)\,\mathbf{1}$. In order to characterize the cubic phase, ψ must have a local minimum at $\mathbf{E} = \mathbf{0}$, and one finds that this requires

$$d_2 < 0, \qquad d_5 > 0, \qquad d_6 > 0. \tag{5.46}$$

Similarly, in order to characterize the tetragonal variants ψ must have a local minimum at $\mathbf{E} = \mathbf{E}_1 = (1/2)(\mathbf{U}_1^2 - \mathbf{I})$, and this necessitates

$$d_4 = \frac{3d_2 + (p-q)d_3}{2(p-q)^2}, \qquad d_7 = -\frac{d_6}{2(2p+q)^2}, \tag{5.47}$$

$$d_5 > 0, \quad 6d_2 - 18d_6 + (p-q)d_3 > 0,$$
$$6d_2 d_6 + (p-q)d_3 d_6 < 0, \quad (p-q)d_3 > 0, \tag{5.48}$$

where we have set

$$p = \tfrac{1}{2}(\eta_1^2 - 1), \qquad q = \tfrac{1}{2}(\eta_3^2 - 1), \tag{5.49}$$

η_1 and η_3 being the lattice parameters.

There are various ways in which one can determine the coefficients $d_2(\theta)$ to $d_7(\theta)$ from experimental measurements. For example, suppose that the elastic moduli of the cubic phase, the lattice parameters of the two phases, and the latent heat of the transformation have been measured. The components of the linear elasticity tensor of the cubic phase (with respect to a cubic basis) are given by

$$\mathrm{C}_{ijkl}^a = \rho \left. \frac{\partial^2 \psi}{\partial E_{ij} \partial E_{kl}} \right|_{\mathbf{E}=0}. \tag{5.50}$$

If $C_{ij}^{(a)}(\theta)$ denote these moduli in Voigt notation, then one can readily show that

$$\rho d_2 = C_{12}^{(a)} - C_{11}^{(a)}, \qquad \rho d_5 = 2C_{44}^{(a)}, \qquad \rho d_6 = \frac{C_{11}^{(a)} + 2C_{12}^{(a)}}{6}. \tag{5.51}$$

We note in passing that the inequalities (5.46), which ensure an energy well at the cubic phase, can be expressed in terms of these moduli by

$$C_{11}^{(a)} - C_{12}^{(a)} > 0, \qquad C_{44}^{(a)} > 0, \qquad C_{11}^{(a)} + 2C_{12}^{(a)} > 0, \tag{5.52}$$

which are the well-known conditions for the positive definiteness of the energy in linear elasticity for a cubic material; see, for example, Ting [9].

Next, the height h between the two energy wells, $h(\theta) = \psi(\mathbf{E}_1, \theta) - \psi(\mathbf{0}, \theta)$, can be calculated from (5.45) to be

$$h(\theta) = \tfrac{1}{54}[-9(p-q)^2 \, d_2(\theta) - (p-q)^3 \, d_3(\theta) + 27(2p+q)^2 \, d_6(\theta)]. \quad (5.53)$$

At the transformation temperature θ_T, we must have $h(\theta_T) = 0$ since the wells then have equal height. The latent heat associated with the transformation from the cubic phase to the tetragonal phase is

$$\lambda = -\left(\psi_\theta\Big|_{\mathbf{E}=\mathbf{E}_1} - \psi_\theta\Big|_{\mathbf{E}=\mathbf{0}}\right)\theta \quad (5.54)$$

and consequently $\lambda(\theta) = -\theta h'(\theta)$. Integrating this and using the fact that $h(\theta_T) = 0$ gives

$$h(\theta) = -\int_{\theta_T}^\theta \frac{\lambda(\theta)}{\theta} d\theta. \quad (5.55)$$

It follows from this and (5.53) that

$$d_3(\theta) = \frac{54}{(p-q)^3}\int_{\theta_T}^\theta \frac{\lambda(\theta)}{\theta}d\theta - \frac{9}{p-q}\,d_2(\theta) + 27\frac{(2p+q)^2}{(p-q)^3}\,d_6(\theta). \quad (5.56)$$

In *summary*, knowledge of the lattice parameters η_1 and η_3, the mass density ρ, the elastic moduli $C_{11}^{(a)}$, $C_{12}^{(a)}$, $C_{44}^{(a)}$ of the cubic phase, and the latent heat λ determines the six coefficients d_2 to d_7 through (5.51), (5.47), and (5.56). James (see Klouček and Luskin [7]) has shown that the response predicted by this generalized form of ψ is in reasonable agreement with the observed behavior of In–Tl.

A free energy function for a material with cubic austenite and *orthorhombic* martensite has been constructed similarly by Vedantam and Abeyaratne [11] and used to characterize a CuAlNi shape memory alloy.

By constructing and analyzing the potential energy function associated with (5.45), one can now determine the stability of the different phases at various stresses and temperatures, and thus construct the corresponding phase diagram in $(\theta, \mathbf{\Sigma})$-space.

REFERENCES

[1] K. Bhattacharya, *Microstructure of Martensite: Why it Forms and How it Gives Rise to the Shape Memory Effect*. Oxford University Press, New York, 2003.

[2] H.B. Callen, *Thermodynamics and an Introduction to Thermostatistics*, second edition. Wiley, New York, 1985.

[3] P. Chadwick, *Continuum Mechanics: Concise Theory and Problems*. Dover, New York, 1999.

[4] J.L. Ericksen, Constitutive theory for some constrained elastic crystals. *International Journal of Solids and Structures*, **22** (1986), pp. 951–64.

[5] A.E. Green and J.E. Adkins, *Large Elastic Deformations*. Clarendon Press, Oxford, 1970.

[6] J. Kestin, *A Course in Thermodynamics*, Volume II. McGraw-Hill, New York, 1979.

[7] P. Klouček and M. Luskin, The computation of the dynamics of the martensitic transformation. *Continuum Mechanics and Thermodynamics*, **6** (1994), pp. 209–40.

[8] G.F. Smith and R.S. Rivlin, The strain energy function for anisotropic elastic materials. *Transactions of the American Mathematical Society*, **88** (1958), pp. 175–93.

[9] T.C.T. Ting, *Anisotropic Elasticity: Theory and Applications*. Oxford University Press, Oxford, 1996.

[10] C. Truesdell and W. Noll, The nonlinear field theories of mechanics, in *Handbuch der Physik,* edited by S. Flügge, Volume III/3. Springer, Berlin, 1965.

[11] S. Vedantam and R. Abeyaratne, A Helmholtz free-energy function for a Cu–Al–Ni shape memory alloy. *International Journal of Nonlinear Mechanics*, **40** (2005), pp. 177–93.

6 The Continuum Theory of Driving Force

6.1 Introduction

In the present chapter, we consider a continuum in which there are three-dimensional, thermomechanical fields involving moving surfaces of discontinuity in strain, particle velocity, and temperature. Our objective is to set out the theory of driving force acting on such surfaces *without specifying any particular constitutive law*. The theory should be applicable to physical settings ranging from the quasistatic response of solids under slow thermal or mechanical loading in which heat conduction is present, to fast adiabatic processes in which temperature need not be continuous and inertia must be taken into account. The theory must be general enough to accommodate such disparate settings.

We begin by stating the fundamental balance laws for momentum and energy in their global form. These laws are then localized where the fields are smooth, leading to the basic field equations of the theory. Localization of the global laws at points where the thermomechanical fields suffer jump discontinuities provides the jump conditions appropriate to such discontinuities.

Without making constitutive assumptions, we then introduce the notion of driving force[1]. The driving force arises through consideration of the entropy production rate associated with the thermomechanical fields under study; it leads to a succinct statement of the implication of the second law of thermodynamics for moving surfaces of discontinuity.

The theory is developed for three space dimensions in *Lagrangian*, or *material*, form, according to which one follows the evolution of fields attached to a given particle of the continuum. The notion of driving force in the continuum theory of phase transformations was developed independently by Truskinovsky [11], Heidug and Lehner [6], and Abeyaratne and Knowles [1, 2]; see also Knowles [7]. The description below is based on [1, 2]. For a discussion of driving force in the materials science literature, see, for example, Christian [4] or Porter and Easterling [8].

---

[1] In earlier publications, we have used the term "driving traction" for the driving force, since it has the units of force per unit area. We now prefer the term "driving force" so that our terminology conforms to that in materials science.

6.2 Balance laws, field equations and jump conditions

6.2.1 Balances of momentum and energy in integral form.
Suppose a body composed of a continuum of unspecified constitutive law occupies a region $\mathcal{R}$ in a reference configuration at a uniform reference temperature. A thermomechanical process for this body involves first a motion, which carries a particle with position vector $\mathbf{x}$ in $\mathcal{R}$ to a point with position vector $\mathbf{y} = \hat{\mathbf{y}}(\mathbf{x}, t)$ at time t. The displacement $\mathbf{u}(\mathbf{x}, t) = \hat{\mathbf{y}}(\mathbf{x}, t) - \mathbf{x}$ is assumed to be continuous in $(\mathbf{x}, t)$ with piecewise continuous first and second partial derivatives for all $\mathbf{x}$ in $\mathcal{R}$ for all times t of interest. The deformation gradient tensor and the particle velocity vector associated with the motion are $\mathbf{F} = \text{Grad } \hat{\mathbf{y}}$, $\mathbf{v} = \partial \mathbf{u}/\partial t$, respectively. It is assumed throughout that $\det \mathbf{F} > 0$.

The process also involves a nominal – or first Piola–Kirchhoff – stress tensor $\sigma(\mathbf{x}, t)$ and the body force per unit mass $\mathbf{b}(\mathbf{x}, t)$ at time t; $\sigma(\cdot, t)$ is required to be piecewise continuous with piecewise continuous gradient on $\mathcal{R}$, while $\mathbf{b}(\cdot, t)$ is assumed continuous there. If ρ is the mass density in the reference configuration, always taken here to be independent of $\mathbf{x}$ for simplicity, the balance laws for linear and angular momenta are then

$$\int_{\partial \mathcal{D}} \sigma \mathbf{n} \, dA + \int_{\mathcal{D}} \rho \mathbf{b} \, dV = \frac{d}{dt} \int_{\mathcal{D}} \rho \mathbf{v} \, dV, \tag{6.1}$$

$$\int_{\partial \mathcal{D}} \hat{\mathbf{y}} \times (\sigma \mathbf{n}) \, dA + \int_{\mathcal{D}} \hat{\mathbf{y}} \times \rho \mathbf{b} \, dV = \frac{d}{dt} \int_{\mathcal{D}} \hat{\mathbf{y}} \times \rho \mathbf{v} \, dV, \tag{6.2}$$

for every bounded subregion $\mathcal{D}$ of $\mathcal{R}$ and for all times. Here $\mathbf{n}$ is the unit outward normal vector on $\partial \mathcal{D}$.

The thermomechanical process also involves a heat flux vector $\mathbf{q}(\mathbf{x}, t)$ (measured per unit referential surface area), a heat supply $r(\mathbf{x}, t)$, and an internal energy $\epsilon(\mathbf{x}, t)$, both measured per unit mass. Unconventionally, we take the heat flux vector $\mathbf{q}$ to be such that $\mathbf{q} \cdot \mathbf{n} > 0$ if heat flows *into* a body across its bounding surface on which $\mathbf{n}$ is the *outward* unit normal vector. Moreover, $\mathbf{q}(\cdot, t)$ is assumed to be piecewise continuous with piecewise continuous gradient on $\mathcal{R}$, while $r(\cdot, t)$ is continuous there; $\epsilon(\mathbf{x}, t)$ is continuous in $(\mathbf{x}, t)$ with piecewise continuous first and second derivatives. The first law of thermodynamics then requires that

$$\int_{\partial \mathcal{D}} \sigma \mathbf{n} \cdot \mathbf{v} \, dA + \int_{\mathcal{D}} \rho \mathbf{b} \cdot \mathbf{v} \, dV + \int_{\partial \mathcal{D}} \mathbf{q} \cdot \mathbf{n} \, dA + \int_{\mathcal{D}} \rho r \, dV$$
$$= \frac{d}{dt} \int_{\mathcal{D}} (\rho \epsilon + \rho \mathbf{v} \cdot \mathbf{v}/2) \, dV. \tag{6.3}$$

6.2.2 Localization of the balance laws.
At any point $\mathbf{x}$ in $\mathcal{R}$ where the fields entering (6.1)–(6.3) are smooth, localization yields the field equations associated with

the balance laws:

$$\text{Div}\,\sigma + \rho\mathbf{b} = \rho\dot{\mathbf{v}} \quad \text{(linear momentum)}, \tag{6.4}$$

$$\sigma\mathbf{F}^T = \mathbf{F}\sigma^T \quad \text{(angular momentum)}, \tag{6.5}$$

$$\sigma \cdot \dot{\mathbf{F}} + \text{Div}\,\mathbf{q} + \rho r = \rho\dot{\epsilon} \quad \text{(energy)}; \tag{6.6}$$

in (6.6), we have used the notation $\mathbf{A} \cdot \mathbf{B}$ to stand for the scalar product of two tensors: $\mathbf{A} \cdot \mathbf{B} = \text{tr}(\mathbf{AB}^T)$.

Next, suppose there is a moving surface S_t in the reference region $\mathcal{R}$ across which $\mathbf{F}$, $\dot{\mathbf{F}}$, $\mathbf{v}$, $\mathbf{q}$, σ, and ϵ suffer jump discontinuities. Note that, since S_t moves *in the reference configuration*, its "laboratory image" – or Eulerian counterpart – does not always separate the same particles. The surface S_t might, for example, be the Lagrangian "pre-image" of a shock wave. In our applications of the theory, we shall find that S_t may sometimes be the Lagrangian interface, or *phase boundary*, between two material phases. We choose a positive side of S_t arbitrarily, and let $\mathbf{n}$ be the unit normal vector pointing into this positive side. Denote by V_n the velocity of propagation of the surface S_t in the direction of $\mathbf{n}$; note that V_n may have either sign. Now suppose $p(\mathbf{x}, t)$ is any Lagrangian field quantity that suffers a jump across S_t, and let p^+ or p^- represent, at a fixed time t, the limiting values of p as a point on S_t is approached from the positive or negative side, respectively. We write

$$[\![p]\!] = p^+ - p^-, \qquad \langle p \rangle = \tfrac{1}{2}(p^+ + p^-) \tag{6.7}$$

for the jump and the average of the values of g on either side of S_t.

From the continuity and piecewise smoothness of $\hat{\mathbf{y}}(\cdot, t)$, it may be shown that

$$[\![\mathbf{F}]\!]\ell = \mathbf{0}, \qquad ([\![\mathbf{F}]\!]\mathbf{n})\,V_n + [\![\mathbf{v}]\!] = \mathbf{0}, \tag{6.8}$$

where ℓ is any vector tangent to S_t. From $(6.8)_1$, it follows that $\mathbf{F}^+$ and $\mathbf{F}^-$ are rank-1 connected:

$$[\![\mathbf{F}]\!] = \mathbf{F}^+ - \mathbf{F}^- = \mathbf{a} \otimes \mathbf{n}, \qquad \text{where} \quad \mathbf{a} = [\![\mathbf{F}]\!]\mathbf{n}. \tag{6.9}$$

When localized at a point on S_t, the balance laws (6.1)–(6.3) yield the following jump conditions:

$$[\![\sigma\mathbf{n}]\!] = -\rho[\![\mathbf{v}]\!]V_n \quad \text{(momentum)} \tag{6.10}$$

$$[\![(\sigma\mathbf{n}) \cdot \mathbf{v}]\!] + [\![\mathbf{q} \cdot \mathbf{n}]\!] = -\rho[\![\epsilon + \tfrac{1}{2}\mathbf{v} \cdot \mathbf{v}]\!]V_n \quad \text{(energy)}. \tag{6.11}$$

The energy jump condition (6.11) can be written in an alternate but equivalent form. By first noting that, for any two vector fields $\mathbf{g}$ and $\mathbf{h}$ that jump across S_t, one has $[\![\mathbf{g} \cdot \mathbf{h}]\!] = \langle\mathbf{g}\rangle \cdot [\![\mathbf{h}]\!] + [\![\mathbf{g}]\!] \cdot \langle\mathbf{h}\rangle$, one can use (6.8), (6.10), and (6.9) to show that

$$\langle(\sigma\mathbf{n}) \cdot \mathbf{v}\rangle = -\langle\sigma\rangle \cdot [\![\mathbf{F}]\!]\,V_n - \rho[\![\mathbf{v} \cdot \mathbf{v}/2]\!]\,V_n. \tag{6.12}$$

Using (6.12) in the energy jump condition, (6.11) gives

$$(\rho[\![\epsilon]\!] - \langle\sigma\rangle \cdot [\![\mathbf{F}]\!])\,V_n + [\![\mathbf{q} \cdot \mathbf{n}]\!] = 0 \quad \text{(energy)}. \tag{6.13}$$

Although our formulation of problems will always be carried out in the Lagrangian framework in use here, we shall at times need to calculate the *true* – or *Cauchy* – stress tensor τ, having determined the nominal – or first Piola–Kirchhoff – stress tensor σ and the deformation gradient tensor $\mathbf{F}$. When this is necessary, we make use of the relation

$$\tau = (1/\det \mathbf{F}) \, \sigma \mathbf{F}^T; \tag{6.14}$$

for derivation and discussion of this basic result in continuum mechanics, the reader may consult Chapter 3 of [3] or Chapter IX of [5] .

6.3 The second law of thermodynamics and the driving force

6.3.1 Entropy production rate.
The thermomechanical process under considera-tion also involves two additional scalar fields, which are essential in addressing the second law of thermodynamics. These are the absolute temperature $\theta(\mathbf{x}, t)$, as-sumed to be positive, and the entropy per unit mass – or specific entropy – $\eta(\mathbf{x}, t)$. It is assumed that $\theta(\cdot, t)$ is piecewise continuous with piecewise continuous gradi-ent on $\mathcal{R}$, while η is piecewise continuous in $(\mathbf{x}, t)$ with piecewise continuous first derivatives. In particular, θ and η may jump across the surface of discontinuity $\mathcal{S}_t$ discussed above.

The rate of entropy production $\Gamma = \Gamma_{\mathcal{D}}(t)$ associated with a bounded subregion $\mathcal{D}$ of $\mathcal{R}$ is defined to be the excess of the rate of increase of the total entropy in $\mathcal{D}$ over the rate that entropy is supplied to $\mathcal{D}$ through $\mathbf{q}$ and r. Thus

$$\Gamma_{\mathcal{D}}(t) = \frac{d}{dt} \int_{\mathcal{D}} \rho \eta \, dV - \int_{\partial \mathcal{D}} \frac{\mathbf{q} \cdot \mathbf{n}}{\theta} \, dA - \int_{\mathcal{D}} \frac{\rho r}{\theta} \, dV. \tag{6.15}$$

Suppose that the surface of discontinuity $\mathcal{S}_t$ intersects $\mathcal{D}$ for some time interval of interest, and that the fields are continuous elsewhere in $\mathcal{D}$. An alternate represen-tation for $\Gamma_{\mathcal{D}}$ may be obtained from (6.15) with the help of a standard calculation that uses the divergence theorem and accounts for the discontinuity on $\mathcal{S}_t$ when carrying out the indicated differentiation; see, for example, p. 116 of [3]. The result may be written in the form

$$\Gamma_{\mathcal{D}} = \Gamma_{\mathcal{D}}^{\text{bulk}} + \Gamma_{\mathcal{D}}^{\text{jump}}, \tag{6.16}$$

where the contribution $\Gamma_{\mathcal{D}}^{\text{bulk}}$ due to particles in the bulk of $\mathcal{D}$ is defined by

$$\Gamma_{\mathcal{D}}^{\text{bulk}} = \int_{\mathcal{D}} \{\rho \dot{\eta} - \text{Div}(\mathbf{q}/\theta) - \rho r/\theta\} \, dV, \tag{6.17}$$

while the entropy production rate $\Gamma_{\mathcal{D}}^{\text{jump}}$ due to the portion of the surface of discon-tinuity that lies in $\mathcal{D}$ is

$$\Gamma_{\mathcal{D}}^{\text{jump}} = -\int_{\mathcal{S}_t \cap \mathcal{D}} \{\rho [\![\eta]\!] \, V_n + [\![\mathbf{q} \cdot \mathbf{n}/\theta]\!]\} \, dA. \tag{6.18}$$

In (6.17), the overdot on η stands for the time derivative at fixed $\mathbf{x}$.

By expanding the divergence in (6.17) and using the local energy balance field equation (6.6), one can find the following alternate form for $\Gamma_{\mathcal{D}}^{\text{bulk}}$:

$$\Gamma_{\mathcal{D}}^{\text{bulk}} = \Gamma_{\mathcal{D}}^{\text{loc}} + \Gamma_{\mathcal{D}}^{\text{con}}, \tag{6.19}$$

where

$$\Gamma_{\mathcal{D}}^{\text{loc}} = \int_{\mathcal{D}} \frac{\rho}{\theta} \left(\frac{1}{\rho} \boldsymbol{\sigma} \cdot \dot{\mathbf{F}} + \theta \dot{\eta} - \dot{\epsilon} \right) dV \tag{6.20}$$

and

$$\Gamma_{\mathcal{D}}^{\text{con}} = \int_{\mathcal{D}} \frac{1}{\theta^2} \mathbf{q} \cdot \nabla \theta \, dV. \tag{6.21}$$

In the decomposition (6.19), $\Gamma_{\mathcal{D}}^{\text{loc}}$ is the aggregate rate of spontaneous production of entropy at the individual particles of $\mathcal{D}$, while $\Gamma_{\mathcal{D}}^{\text{con}}$ stands for the rate of entropy production due to heat conduction.

The Helmholtz free energy $\psi(\mathbf{x}, t)$ per unit mass at time t at the particle labeled by $\mathbf{x}$ is defined by

$$\psi = \epsilon - \theta\eta. \tag{6.22}$$

In terms of ψ, one may rewrite $\Gamma_{\mathcal{D}}^{\text{loc}}$ as follows:

$$\Gamma_{\mathcal{D}}^{\text{loc}} = \int_{\mathcal{D}} \frac{\rho}{\theta} \left(\frac{1}{\rho} \boldsymbol{\sigma} \cdot \dot{\mathbf{F}} - \eta \dot{\theta} - \dot{\psi} \right) dV. \tag{6.23}$$

In the setting of smooth thermomechanical processes, the decomposition (6.19) may be found in Section 79 of Truesdell and Noll [9]; see also the discussion in Chapter 2 of [10].

By using the alternative energy jump condition (6.13), one can rewrite part of the integrand in (6.18) as follows:

$$[\![\mathbf{q} \cdot \mathbf{n}/\theta]\!] = \left(-\frac{\rho[\![\epsilon]\!] - \langle\boldsymbol{\sigma}\rangle \cdot [\![\mathbf{F}]\!]}{\langle\theta\rangle} \right) V_n \tag{6.24}$$

$$+ (\langle 1/\theta \rangle - 1/\langle\theta\rangle) [\![\mathbf{q} \cdot \mathbf{n}]\!] + [\![1/\theta]\!]\langle\mathbf{q} \cdot \mathbf{n}\rangle.$$

In this book, the thermomechanical processes to be considered will either be adiabatic, in which case $\mathbf{q} = \mathbf{0}, r = 0$, or will involve heat conduction and the temperature will be continuous. For adiabatic processes the last two terms on the right in (6.24) vanish by virtue of the vanishing of $\mathbf{q}$. For processes involving heat conduction, these same terms vanish because of the continuity of the temperature. Thus in all thermomechanical processes that we shall encounter, (6.24) reduces to

$$[\![\mathbf{q} \cdot \mathbf{n}/\theta]\!] = -\frac{1}{\langle\theta\rangle} (\rho[\![\epsilon]\!] - \boldsymbol{\sigma} \cdot [\![\mathbf{F}]\!]) V_n, \tag{6.25}$$

so that the contribution $\Gamma_{\mathcal{D}}^{\text{jump}}$ of (6.18) to the rate of entropy production may be rewritten as

$$\Gamma_{\mathcal{D}}^{\text{jump}} = \int_{\mathcal{S}_t \cap \mathcal{D}} \frac{[\![\rho\epsilon]\!] - \langle\boldsymbol{\sigma}\rangle \cdot [\![\mathbf{F}]\!] - \langle\theta\rangle[\![\rho\eta]\!]}{\langle\theta\rangle} V_n \, dA. \qquad (6.26)$$

In terms of the Helmholtz free energy ψ, (6.26) becomes

$$\Gamma_{\mathcal{D}}^{\text{jump}} = \int_{\mathcal{S}_t \cap \mathcal{D}} \frac{[\![\rho\psi]\!] - \langle\boldsymbol{\sigma}\rangle \cdot [\![\mathbf{F}]\!] + \langle\rho\eta\rangle[\![\theta]\!]}{\langle\theta\rangle} V_n \, dA. \qquad (6.27)$$

6.3.2 Driving force and the second law. Motivated by (6.26) and (6.27), we define the *driving force* $f(\mathbf{x}, t)$ acting at a point on the surface of discontinuity $\mathcal{S}_t$ by either of the following equivalent formulas:

$$f = [\![\rho\psi]\!] - \langle\boldsymbol{\sigma}\rangle \cdot [\![\mathbf{F}]\!] + \langle\rho\eta\rangle[\![\theta]\!] = [\![\rho\epsilon]\!] - \langle\boldsymbol{\sigma}\rangle \cdot [\![\mathbf{F}]\!] - \langle\theta\rangle[\![\rho\eta]\!], \quad (6.28)$$

[1, 2]. By (6.27) and (6.28), the contribution $\Gamma_{\mathcal{D}}^{\text{jump}}$ to the entropy production rate for $\mathcal{D}$ may then be written as

$$\Gamma_{\mathcal{D}}^{\text{jump}} = \int_{\mathcal{D} \cap \mathcal{S}_t} \frac{f}{\langle\theta\rangle} V_n \, dA. \qquad (6.29)$$

In its global form, the second law of thermodynamics requires the Clausius–Duhem inequality to hold:

$$\Gamma_{\mathcal{D}}(t) \geq 0 \qquad (6.30)$$

for all times and for all bounded subregions $\mathcal{D}$ of $\mathcal{R}$. By first localizing (6.30) at a point $\mathbf{x}$ at which the fields are smooth, one finds that

$$\rho\dot{\eta} - \text{Div}(\mathbf{q}/\theta) - \rho r/\theta \geq 0 \quad \text{(entropy inequality in bulk)} \quad (6.31)$$

at all such points of smoothness. It then follows that

$$\Gamma_{\mathcal{D}}^{\text{bulk}} \geq 0 \qquad (6.32)$$

for every bounded $\mathcal{D}$ and for all times.

On the other hand, localizing (6.30) at a point on the surface $\mathcal{S}_t$ and using (6.16) and (6.17) leads to the additional requirement

$$f \, V_n \geq 0 \quad \text{(entropy inequality on surface of discontinuity)} \quad (6.33)$$

at all points on $\mathcal{S}_t$. Thus according to the second law, the sign of V_n, and therefore the direction of advance of the surface $\mathcal{S}_t$, is determined by the sign of the driving force f acting on $\mathcal{S}_t$. A result precisely analogous to (6.33) was found to hold true in the purely mechanical theory of Chapter 4; see (4.10).

Suppose the thermomechanical process is adiabatic, so that $\mathbf{q} = \mathbf{0}, r = 0$. Then by (6.13), one has

$$([\![\rho\epsilon]\!] - \langle\boldsymbol{\sigma}\rangle \cdot [\![\mathbf{F}]\!]) V_n = 0, \qquad (6.34)$$

so that from $(6.28)_2$ the driving force becomes

$$f = -\langle\theta\rangle[\![\rho\eta]\!], \quad \text{(driving force – adiabatic)} \tag{6.35}$$

and the entropy inequality (6.33) reduces to

$$[\![\eta]\!] \, V_n \leq 0. \tag{6.36}$$

Thus in an adiabatic process, the specific entropy of a particle cannot decrease as the surface S_t sweeps past, a result familiar from inviscid gas dynamics.

For a nonadiabatic process, continuity of temperature and $(6.28)_1$ imply that

$$f = [\![\rho\psi]\!] - \langle\sigma\rangle \cdot [\![\mathbf{F}]\!] \quad \text{(driving force – nonadiabatic).} \tag{6.37}$$

6.3.3 Driving force in the case of mechanical equilibrium.

Suppose that *mechanical* equilibrium prevails, in the sense that the right-hand sides of the two consequences (6.4) and (6.10) of momentum balance are replaced by null vectors, and assume that the temperature is continuous everywhere. Equation (6.10) now asserts that traction is continuous across the surface S_t of strain discontinuity: $[\![\sigma\mathbf{n}]\!] = \mathbf{0}$. With the help of the rank-one connection (6.9), one then finds that

$$\langle\sigma\rangle \cdot [\![\mathbf{F}]\!] = \langle\sigma\mathbf{n}\rangle \cdot \mathbf{a} = (\sigma^{\pm}\mathbf{n}) \cdot \mathbf{a} = \sigma^{\pm} \cdot [\![\mathbf{F}]\!]. \tag{6.38}$$

This result, together with continuity of temperature, may then be used to specialize the representation $(6.28)_1$ of the driving force acting on a strain discontinuity to the case of mechanical equilibrium, yielding

$$f = [\![\rho\psi]\!] - (\sigma^{\pm}\mathbf{n}) \cdot \mathbf{a} = [\![\rho\psi]\!] - \sigma^{\pm} \cdot [\![\mathbf{F}]\!] \quad \begin{array}{l}\text{(driving force –}\\ \text{mechanical equilibrium).}\end{array} \tag{6.39}$$

It may be remarked that the version (6.39) of driving force appropriate to the case of mechanical equilibrium is not in general simply related to the "resolved" Piola–Kirchhoff shear stress $(\mathbf{I} - \mathbf{n} \otimes \mathbf{n})\sigma\mathbf{n}$. Moreover, when the contributions due to inertia are *retained* in (6.4) and (6.10), the notion of resolved shear stress on S_t is itself ambiguous, since the nominal traction $\sigma\mathbf{n}$ need no longer be continuous across the discontinuity, so that resolved shear stress will in general be different on the two sides of S_t.

REFERENCES

[1] R. Abeyaratne and J.K. Knowles, On the driving traction acting on a surface of strain discontinuity in a continuum. *Journal of the Mechanics and Physics of Solids*, **38** (1990), pp. 345–60.

[2] R Abeyaratne and J.K. Knowles, A note on the driving traction acting on a propagating interface: adiabatic and non-adiabatic processes of a continuum. *ASME Journal of Applied Mechanics*, **67** (2000), pp. 829–31.

[3] P. Chadwick, *Continuum Mechanics*. Dover, New York, 1999.

[4] J.W. Christian, *The Theory of Martensitic Transformations in Metals and Alloys*, Part 1. Pergamon, Oxford, 1975.

[5] M.E. Gurtin, *An Introduction to Continuum Mechanics*. Academic Press, New York, 1981.

[6] W. Heidug and F.K. Lehner, Thermodynamics of coherent phase transformations in non-hydrostatically stressed solids. *Pure and Applied Geophysics*, **123** (1985), pp. 91–8.

[7] J.K. Knowles, On the dissipation associated with equilibrium shocks in finite elasticity. *Journal of Elasticity*, **9** (1979), pp. 131–58.

[8] D.A. Porter and K.E. Easterling, *Phase Transformations in Metals and Alloys*. Van Nostrand-Reinhold (UK) Ltd, London, 1981.

[9] C. Truesdell and W. Noll, The non-linear field theories of mechanics, in *Handbuch der Physik* III/3, edited by S. Flügge. Springer, Berlin, 1965.

[10] C. Truesdell, *Rational Thermodynamics*. McGraw-Hill, New York, 1969.

[11] L. Truskinovsky, Equilibrium phase interfaces. *Soviet Physics Doklady*, **27** (1982), pp. 551–3.

7 Thermoelastic Materials

7.1 Introduction

The developments in the preceding chapter assumed nothing about the constitutive response of the continuum; we now restrict attention to a special class of constitutive laws – the so-called thermoelastic materials introduced previously in Chapter 5. Our discussion in Chapter 5 was focused entirely on the energy wells of the characterizing energy potential. Here we discuss thermoelastic materials and nonlinear thermoelasticity in more detail.

In Section 7.2 we state the constitutive law of nonlinear thermoelasticity, in which stress and specific entropy are specified as functions of deformation gradient and absolute temperature through the Helmholtz free energy potential. An equivalent alternate form of the constitutive law, in which stress and temperature are given in terms of deformation gradient and specific entropy by means of the internal energy potential, is also discussed. The expression for the driving force is then specialized to this setting. Next we state the heat conduction law, and in Section 7.2.3, we write out the full theory in the form of four scalar partial differential equations involving the three components of displacement and temperature. The accompanying jump conditions are also laid out. In the final subsection we specialize the results to a state of thermomechanical equilibrium.

Once the full theory of thermoelasticity has been assembled, in Section 7.3 we summarize some results concerning the notion of *material stability* based on an analysis, given elsewhere, of the linearized dynamic stability of a body that has been homogeneously deformed from a given reference configuration at a given temperature to a new configuration at a new temperature.

Finally, we specialize the theory to the one-dimensional kinematics of *uniaxial strain* in Section 7.4.

7.2 The thermoelastic constitutive law

7.2.1 Relations among stress, deformation gradient, temperature, and specific entropy. As introduced in Chapter 5, a material is said to be *thermoelastic* if there

is a *Helmholtz free energy potential* $\hat{\psi}(\mathbf{F}, \theta)$ such that the Helmholtz free energy field $\psi(\mathbf{x}, t)$ of the preceding chapter is given by $\hat{\psi}(\mathbf{F}(\mathbf{x}, t), \theta(\mathbf{x}, t))$, and the nominal stress tensor and the specific entropy are related to $\mathbf{F}$ and θ by

$$\boldsymbol{\sigma} = \hat{\boldsymbol{\sigma}}(\mathbf{F}, \theta) = \rho\hat{\psi}_\mathbf{F}(\mathbf{F}, \theta), \qquad \eta = \hat{\eta}(\mathbf{F}, \theta) = -\hat{\psi}_\theta(\mathbf{F}, \theta), \qquad (7.1)$$

where, as before, ρ is the referential mass density, and the subscripts attached to $\hat{\psi}$ indicate the gradient with respect to $\mathbf{F}$ and the partial derivative with respect to θ.

As discussed in Chapter 5, at each fixed temperature θ, the value $\hat{\psi}(\mathbf{F}, \theta)$ of the Helmholtz potential should be invariant under changes of observer, a requirement that is sometimes called the principle of objectivity or of material frame indifference. This leads to the conclusion that $\hat{\psi}(\mathbf{F}, \theta)$ must depend on $\mathbf{F}$ only through the stretch tensor $\mathbf{U}$ or the so-called *right Cauchy–Green tensor* $\mathbf{C} = \mathbf{F}^T\mathbf{F} = \mathbf{U}^2$. The proof of this result for *thermoelasticity* is identical with that for pure elasticity; see Chapter IX, including Exercise 1, of [3]. It is easy to show that this special dependence of $\hat{\psi}(\mathbf{F}, \theta)$ on $\mathbf{F}$ implies that the local balance of angular momentum (6.5) is automatically satisfied for a thermoelastic material, and therefore that the true stress tensor $\boldsymbol{\tau}$ of (6.14) is automatically symmetric: $\boldsymbol{\tau} = \boldsymbol{\tau}^T$. Since we shall always assume that $\hat{\psi}(\mathbf{F}, \theta)$ is invariant under changes of observer, we shall omit (6.5) from the list of field equations for thermoelastic materials.

If at each temperature the material is isotropic in the reference state, then there is a further simplification in the dependence of $\hat{\psi}(\mathbf{F}, \theta)$ on $\mathbf{F}$:

$$\hat{\psi}(\mathbf{F}, \theta) = \bar{\psi}(I_1, I_2, I_3, \theta), \qquad (7.2)$$

where the *fundamental scalar invariants* $I_j(\mathbf{C})$ are given by

$$I_1(\mathbf{C}) = \text{tr } \mathbf{C}, \quad I_2(\mathbf{C}) = \tfrac{1}{2}\left[(\text{tr } \mathbf{C})^2 - \text{tr } (\mathbf{C}^2)\right], \quad I_3(\mathbf{C}) = \det \mathbf{C}. \qquad (7.3)$$

The *specific heat at constant deformation gradient* is defined to be

$$c(\mathbf{F}, \theta) = -\theta \,\hat{\psi}_{\theta\theta}(\mathbf{F}, \theta) = \theta \,\hat{\eta}_\theta(\mathbf{F}, \theta). \qquad (7.4)$$

We shall always assume that $c(\mathbf{F}, \theta)$ is positive. Since the absolute temperature is also positive, $(7.4)_2$ implies $\hat{\eta}(\mathbf{F}, \theta)$ is an increasing function of θ at fixed $\mathbf{F}$. It follows that, for each fixed $\mathbf{F}$, (7.1) can be inverted to express the temperature $\theta = \hat{\theta}(\mathbf{F}, \eta)$ in terms of $\mathbf{F}$ and η. As a result, one may define an *internal energy potential* through

$$\hat{\epsilon}(\mathbf{F}, \eta) = \hat{\psi}(\mathbf{F}, \hat{\theta}(\mathbf{F}, \eta)) + \eta\hat{\theta}(\mathbf{F}, \eta); \qquad (7.5)$$

cf. (6.22). It follows from (7.5) that the constitutive law (7.1) can be written in the following alternate form:

$$\boldsymbol{\sigma} = \rho\hat{\epsilon}_\mathbf{F}(\mathbf{F}, \eta), \qquad \theta = \hat{\epsilon}_\eta(\mathbf{F}, \eta); \qquad (7.6)$$

thus the two potentials $\hat{\psi}$ and $\hat{\epsilon}$ are Legendre transforms of one another.

Using (7.5) and (7.6) in the local version (6.6) of the first law leads to the following alternate local energy equation for thermoelastic materials:

$$\text{Div } \mathbf{q} + \rho r = \rho \theta \dot{\eta}. \tag{7.7}$$

Thus in the thermoelastic case, the basic local balance laws for linear momentum and energy may be written in the forms (6.4) and (7.7), respectively.

For a thermoelastic material, it is readily shown that the integrand in (6.23) vanishes identically so that, by (6.19), the various contributions to the entropy production rate satisfy

$$\Gamma_{\mathcal{D}}^{\text{loc}} = 0, \qquad \Gamma_{\mathcal{D}}^{\text{bulk}} = \Gamma_{\mathcal{D}}^{\text{con}}. \tag{7.8}$$

It follows from (6.16) that

$$\Gamma_{\mathcal{D}} = \Gamma_{\mathcal{D}}^{\text{con}} + \Gamma_{\mathcal{D}}^{\text{jump}}. \tag{7.9}$$

Thus thermoelastic materials produce entropy only through heat conduction and the propagation of surfaces of discontinuity. Indeed, if the thermomechanical process under study is adiabatic, so that $\Gamma_{\mathcal{D}}^{\text{con}} = 0$, entropy production arises *solely* through moving discontinuities such as $\mathcal{S}_t$.

Using (7.8) and (6.21) in (6.32) and localizing at a point of smoothness yields

$$\mathbf{q} \cdot \nabla \theta \geq 0. \tag{7.10}$$

When specialized to a thermoelastic material, the two equivalent representations in (6.28) for the driving force f on $\mathcal{S}_t$ yield the following striking formulas:

$$
\begin{aligned}
f/\rho &= [\![\hat{\psi}]\!] - \langle \hat{\psi}_{\mathbf{F}} \rangle \cdot [\![\mathbf{F}]\!] - \langle \hat{\psi}_\theta \rangle [\![\theta]\!] \\
&= [\![\hat{\epsilon}]\!] - \langle \hat{\epsilon}_{\mathbf{F}} \rangle \cdot [\![\mathbf{F}]\!] - \langle \hat{\epsilon}_\eta \rangle [\![\eta]\!].
\end{aligned} \tag{7.11}
$$

If the thermomechanical process is adiabatic, then (6.13) with $\mathbf{q} = \mathbf{0}$ shows that $(7.11)_2$ reduces to

$$f/\rho = -\langle \hat{\epsilon}_\eta \rangle [\![\eta]\!] = -\langle \theta \rangle [\![\eta]\!] \quad \text{(adiabatic);} \tag{7.12}$$

cf. (6.35). For nonadiabatic processes in a thermoelastic material,

$$f/\rho = [\![\hat{\psi}]\!] - \langle \hat{\psi}_{\mathbf{F}} \rangle \cdot [\![\mathbf{F}]\!] \quad \text{(nonadiabatic).} \tag{7.13}$$

Suppose that the thermoelastic body undergoes an isothermal homogeneous motion with gradient $\mathbf{F}(t)$ at time t, for $t_1 \leq t \leq t_2$. Let $\mathbf{F}(t_1) = \mathbf{F}^+$, $\mathbf{F}(t_2) = \mathbf{F}^-$. Consider an arbitrary sub-body and let $\mathcal{D}$ be the region it occupies in the reference configuration. Integration of (7.7) over $\mathcal{D}$, followed by integration of the result with respect to t over (t_1, t_2), leads to the assertion that the total heat released by the sub-body during this isothermal process is given by λM where M is the mass of the sub-body, and the heat released per unit mass λ is given by

$$\lambda = -\theta \left[\eta(\mathbf{F}^-, \theta) - \eta(\mathbf{F}^+, \theta) \right]. \tag{7.14}$$

Equation (7.14) has an obvious local interpretation for isothermal processes that involve nonhomogeneous motions. If $\mathbf{F}^+$ and $\mathbf{F}^-$ correspond to two phases of the material then λ is the *latent heat* per unit mass associated with the phase transformation $(\mathbf{F}^+, \theta) \to (\mathbf{F}^-, \theta)$.

7.2.2 The heat conduction law. For a thermoelastic material we assume that the heat conduction law has the form

$$\mathbf{q} = \mathbf{K}(\mathbf{F}, \theta) \operatorname{Grad} \theta, \tag{7.15}$$

where the referential heat conductivity tensor $\mathbf{K}(\mathbf{F}, \theta)$ is assumed to be symmetric. By (7.10), it follows that

$$(\mathbf{K}(\mathbf{F}, \theta)\mathbf{g}) \cdot \mathbf{g} \geq 0, \tag{7.16}$$

where $\mathbf{g} = \operatorname{Grad} \theta$. If it is assumed that, for an arbitrarily chosen vector $\mathbf{g}$, there is a thermomechanical process sustainable by the body for which, at some particle, $\operatorname{Grad} \theta(\mathbf{x}, t) = \mathbf{g}$, then (7.16) must hold for every $\mathbf{g}$, whence it follows that $\mathbf{K}(\mathbf{F}, \theta)$ is nonnegative definite for every $\mathbf{F}, \theta$, which we henceforth assume.

7.2.3 The partial differential equations of nonlinear thermoelasticity. We now state the basic equations of nonlinear thermoelasticity in terms of the components of the various fields in a fixed rectangular Cartesian coordinate frame, in which, for example, the position vector $\mathbf{x}$ and the deformation gradient tensor $\mathbf{F}$ have the respective components x_i and F_{ij}. The constitutive equations (7.1) and (7.15) take the form

$$\sigma_{ij} = \rho \hat{\psi}_{F_{ij}}(\underline{F}, \theta), \qquad \eta = -\hat{\psi}_\theta(\underline{F}, \theta), \tag{7.17}$$

$$q_i = K_{ij}(\underline{F}, \theta)\theta_{,j}. \tag{7.18}$$

We use the symbol $\underline{A}$ to stand for the matrix of components of the 2-tensor $\mathbf{A}$ in the given frame; the summation convention for repeated subscripts such as $i, j, k, \ldots$ is in use, and a comma preceding a subscript indicates differentiation with respect to the corresponding x-coordinate.

The local forms (6.4) and (7.7) of the balance laws yield

$$\sigma_{ij,j} + \rho b_i = \rho \dot{v}_i, \tag{7.19}$$

$$q_{j,j} + \rho r = \rho \theta \dot{\eta}. \tag{7.20}$$

Using (7.17) and (7.18) in (7.19) and (7.20) and recalling that $\mathbf{F} = \operatorname{Grad} \hat{\mathbf{y}}(\mathbf{x}, t) = \mathbf{1} + \operatorname{Grad} \mathbf{u}(\mathbf{x}, t)$ leads to the fundamental system of partial differential equations for the three components $u_i(\mathbf{x}, t)$ of displacement and the temperature $\theta(\mathbf{x}, t)$:

$$c_{ijkl}(\underline{F}, \theta)u_{k,jl} + m_{ij}(\underline{F}, \theta)\theta_{,j} + \rho b_i - \rho \ddot{u}_i = 0, \tag{7.21}$$

$$\left[K_{ij}(\underline{F}, \theta)\theta_j\right]_{,i} + \rho r + \theta m_{ij}(\underline{F}, \theta)\dot{F}_{ij} - \rho c(\underline{F}, \theta)\dot{\theta} = 0. \tag{7.22}$$

Respectively, c_{ijkl} and m_{ij} are the components of the 4-tensor $\mathbf{C}$ and the 2-tensor $\mathbf{M}$ defined by

$$\mathbf{C} = \mathbf{C}(\mathbf{F}, \theta) = \rho\, \hat{\psi}_{\mathbf{FF}}(\mathbf{F}, \theta), \qquad \mathbf{M} = \mathbf{M}(\mathbf{F}, \theta) = \rho\hat{\psi}_{\mathbf{F}\theta}(\mathbf{F}, \theta); \quad (7.23)$$

$\mathbf{C}(\mathbf{F}, \theta)$ and $\mathbf{M}(\mathbf{F}, \theta)$ are respectively called the *elasticity* tensor and the *stress–temperature tensor*.

In component form, the kinematic jump conditions (6.8) are

$$[\![F_{ij}]\!]l_j = 0, \qquad [\![F_{ij}]\!]n_j V_n + [\![v_i]\!] = 0; \quad (7.24)$$

the associated rank-1 connection relating $\mathbf{F}^{\pm}$ is

$$F_{ij}^+ - F_{ij}^- = a_i n_j. \quad (7.25)$$

The momentum and energy jump conditions (6.10) and (6.11) are

$$[\![\sigma_{ij}]\!]n_j + \rho[\![v_i]\!]V_n = 0, \quad (7.26)$$

$$[\![\sigma_{ij}v_i]\!]n_j + \rho[\![\epsilon + (1/2)v_j v_j]\!]V_n + [\![q_j]\!]n_j = 0. \quad (7.27)$$

The component form of the alternate version (6.13) of the energy jump condition is

$$\rho[\![\epsilon]\!]V_n - \langle\sigma_{ij}\rangle[\![F_{ij}]\!]V_n + [\![q_j]\!]n_j = 0, \quad (7.28)$$

and finally, the representations (7.11) for the driving force may be written as

$$\begin{aligned}
f/\rho &= [\![\hat{\psi}]\!] - \langle\hat{\psi}_{F_{ij}}\rangle[\![F_{ij}]\!] - \langle\hat{\psi}_\theta\rangle[\![\theta]\!] \\
&= [\![\hat{\epsilon}]\!] - \langle\hat{\epsilon}_{F_{ij}}\rangle[\![F_{ij}]\!] - \langle\hat{\epsilon}_\eta\rangle[\![\eta]\!].
\end{aligned} \quad (7.29)$$

7.2.4 Thermomechanical equilibrium.

Consider the case in which the body force $\mathbf{b}$ and the heat supply r vanish, and suppose that (i) the fields $\mathbf{u}$ and θ are independent of time and (ii) the temperature is continuous. Then $\mathbf{v} = \mathbf{0}$, and by the thermoelastic constitutive law (7.1) and the heat conduction law (7.15), all fields are independent of time, so that the field equations (6.4) and (6.6) for linear momentum and energy become

$$\text{Div } \boldsymbol{\sigma} = \mathbf{0} \quad \text{or} \quad \sigma_{ij,j} = 0, \quad (7.30)$$

$$\text{Div } \mathbf{q} = 0 \quad \text{or} \quad q_{j,j} = 0. \quad (7.31)$$

The jump conditions (6.10) and (6.11) specialize to

$$[\![\boldsymbol{\sigma}]\!]\mathbf{n} = \mathbf{0} \quad \text{or} \quad [\![\sigma_{ij}]\!]n_j = 0, \quad (7.32)$$

$$[\![\mathbf{q}]\!] \cdot \mathbf{n} = 0 \quad \text{or} \quad [\![q_j]\!]n_j = 0. \quad (7.33)$$

Clearly mechanical equilibrium holds at present, so we may make use of (6.38) to note that

$$\langle\hat{\psi}_{\mathbf{F}}\rangle \cdot [\![\mathbf{F}]\!] = \hat{\psi}_{\mathbf{F}}^{\pm} \cdot [\![\mathbf{F}]\!]. \quad (7.34)$$

The formula $(7.11)_1$ for driving force may therefore be replaced by

$$f/\rho = [\![\hat{\psi}]\!] - \hat{\psi}_{\mathbf{F}}^{\pm} \cdot [\![\mathbf{F}]\!] = [\![\hat{\psi}]\!] - \hat{\psi}_{F_{ij}}^{\pm}[\![F_{ij}]\!]. \qquad (7.35)$$

Let $\hat{P}(\mathbf{F}, \theta; \sigma)$ be the potential energy per unit volume; see (5.10):

$$\hat{P}(\mathbf{F}, \theta; \sigma) = \rho\hat{\psi}(\mathbf{F}, \theta) - \sigma \cdot \mathbf{F}. \qquad (7.36)$$

For the case of thermomechanical equilibrium under discussion, it is clear from (7.35) and (7.36) that the driving force coincides with the jump in potential energy per unit volume, in the sense that

$$f = [\![\hat{P}(\mathbf{F}, \theta; \rho\hat{\psi}_{\mathbf{F}}^{\pm}(\mathbf{F}, \theta))]\!]. \qquad (7.37)$$

7.3 Stability of a thermoelastic material

For background purposes, we discuss here the notion of *material stability* in the setting of thermoelasticity. Since the results presented in this section are not explicitly needed in what follows, we only state them, citing references for further information. In particular, detailed proofs of the principal results may be found in [1].

For an elastic – as distinguished from thermoelastic – body, Hadamard related the stability of an equilibrium configuration subject to prescribed displacements on its boundary to the existence of real speeds of propagation of discontinuities for the same material; see Sections 269–71 of [4]. Such an equilibrium state was viewed in [4] as stable if the associated stored elastic energy has a relative minimum within the class of configurations of the body that satisfy the specified boundary conditions. The fact that the second variation of the stored energy must be nonnegative at such a stable equilibrium configuration led Hadamard to a local necessary condition that is identical to that which guarantees the existence of real wave speeds for acceleration jumps propagating in an arbitrary direction. The latter condition asserts that the local acoustic tensor of the elastic medium must be nonnegative definite for all directions of propagation of the wave. For a more modern discussion and generalization of Hadamard's results, see Sections 68*bis*, 71, and 89 of [9].

A condition closely related to the one found by Hadamard is that of *strong ellipticity* of the elastic material under study. An elasticity tensor corresponding to a given configuration is said to be strongly elliptic if the associated acoustic tensor is strictly positive definite for all directions of propagation of small-amplitude, plane, sinusoidal waves superposed on the underlying configuration; see Section 44 of [9] and Section 25 of [2]. In a one-dimensional setting such as that for uniaxial strain developed earlier in Chapter 2, strong ellipticity at the given state of uniaxial strain reduces to the assertion that the slope of the stress–strain curve at that state be positive. In three dimensions, there is no such simple physical interpretation of the strong ellipticity condition. Elastic materials that *lose* strong ellipticity at certain configurations play a major role in the theory of solids that may develop microstructure associated with stress-induced phase transitions; see, for example, [5, 6].

In particular, loss of strong ellipticity at *some* configuration is *necessary* for the existence of a stress-induced equilibrium mixture of phases; see [7]. In a one-dimensional theory, the simplest such "nonelliptic" material is one whose stress–strain curve has a declining branch that separates two rising branches, as did the examples in Chapters 2 and 3.

An elastic material is said to be *stable* if its stored energy function – or elastic potential – satisfies the strong ellipticity condition in *all* configurations. Since stress-induced phase transitions are extensively studied, elastic materials that are *not* stable are not to be rejected as pathological.

Our intent here is to define a similar notion of stability for a *thermoelastic* material, and to state, without proof, conditions on the Helmholtz free energy potential that, taken together, provide the counterpart of the strong ellipticity condition for an elastic material in the sense that they are necessary and sufficient for material stability. Our approach is based on the study of small amplitude sinusoidal waves superposed on a homogeneously deformed thermoelastic body at a uniform temperature. The respective perturbations $\mathbf{u}(\mathbf{x}, t)$ in displacement and $T(\mathbf{x}, t)$ in temperature associated with such waves satisfy the differential equations obtained by linearizing the field equations (7.21)–(7.23) about $\mathbf{F}^0$, θ^0, where θ^0 is the temperature of the base state, and $\mathbf{F}^0$ is the deformation gradient associated with the underlying homogeneous deformation from the reference configuration to the configuration of the base state. Assuming these perturbations to have the form $\mathbf{u}(\mathbf{x}, t) = \exp(ik\mathbf{n} \cdot \mathbf{x})\mathbf{g}(t)$, $T(\mathbf{x}, t) = \exp(ik\mathbf{n} \cdot \mathbf{x})\phi(t)$ leads to linear ordinary differential equations for $\mathbf{g}(t)$ and $\phi(t)$, which involve $\mathbf{F}^0$ and θ^0. We seek necessary and sufficient conditions on the material for $\mathbf{g}(t)$ and $\phi(t)$ to remain bounded as $t \to \infty$ for all wave numbers $k > 0$ and all unit vectors $\mathbf{n}$; a thermoelastic material that satisfies these conditions for all permissible base states $\mathbf{F}^0, \theta^0$ will be called stable.

With a given thermoelastic material, one may associate both *isothermal* and *isentropic* acoustic tensors. Both are 2-tensor-valued functions of a unit vector $\mathbf{n}$ that specifies the direction of propagation of an infinitesimal wave, and both depend on the underlying base state through $\mathbf{F}^0$ and θ^0. The isothermal acoustic tensor $\mathbf{A}(\mathbf{n})$, based on the Helmholtz free energy, is defined through its components in an orthonormal basis by

$$A_{ik}(\mathbf{n}) = (1/\rho)c_{ijkl}(\mathbf{F}_0, \theta_0)n_j n_l, \tag{7.38}$$

where $c_{ijkl}(\mathbf{F}^0, \theta^0)$ are the components of the elasticity tensor at the base state (see $(7.23)_1$), and ρ is the referential mass density.

The isentropic acoustic tensor $\mathbf{H}(\mathbf{n})$ is defined by a formula like (7.38), but with the $c_{ijkl}(\mathbf{F}, \theta)$ replaced by the components of a 4-tensor defined as in $(7.23)_1$, except that $\hat{\psi}_{\mathbf{FF}}(\mathbf{F}, \theta)$ is replaced by $\hat{\epsilon}_{\mathbf{FF}}(\mathbf{F}, \eta)$; these components are then evaluated at $\mathbf{F}^0, \eta^0$, where η^0 is the specific entropy at the base state. Thus the isentropic acoustic tensor bears the same formal relation to the internal energy potential at fixed η that the isothermal acoustic tensor bears to the Helmholtz free energy potential at fixed θ. One can then show that

$$\mathbf{H}(\mathbf{n}) = \mathbf{A}(\mathbf{n}) + (\theta^0/c)\,\mathbf{p}(\mathbf{n}) \otimes \mathbf{p}(\mathbf{n}). \qquad (7.39)$$

Here $c = c(\mathbf{F}^0, \theta^0)$ is the *specific heat at constant deformation gradient* at the base state defined through (7.4), and $\mathbf{p}(\mathbf{n})$ is a vector-valued function of $\mathbf{n}$ given by

$$\mathbf{p}(\mathbf{n}) = (1/\rho)\,\mathbf{M}^0\mathbf{n}, \qquad (7.40)$$

$\mathbf{M}^0 = \mathbf{M}(\mathbf{F}^0, \theta^0)$ being the base-state value of the stress–temperature tensor defined in $(7.23)_2$.

In the adiabatic theory, for which the heat flux vector $\mathbf{q}$ and the heat supply r vanish in all processes, one finds that a necessary and sufficient condition for material stability is the positive definiteness of the isentropic acoustic tensor for all propagation directions $\mathbf{n}$ at all base states $\mathbf{F}^0, \theta^0$. From (7.39), one notes that positive definiteness of the *isothermal* acoustic tensor is sufficient, but not necessary, for positive definiteness of the isentropic acoustic tensor.

Let $a_j(\mathbf{n})$ be the (necessarily real) eigenvalues of the symmetric isothermal acoustic tensor $\mathbf{A}(\mathbf{n})$, with $\mathbf{e}_j(\mathbf{n})$ the corresponding orthonormal eigenvectors. Let $p_j(\mathbf{n}) = \mathbf{p}(\mathbf{n}) \cdot \mathbf{e}_j$ be the $\mathbf{e}_j$-component of $\mathbf{p}(\mathbf{n})$. When heat conduction is present, a thermoelastic material is stable if and only if the following three conditions all hold:

$$\left. \begin{array}{ll} c > 0, & \\[2mm] a_j(\mathbf{n}) \geq 0 & \text{for every } \mathbf{n},\ j = 1,2,3, \\[2mm] a_j^2(\mathbf{n}) + p_j^2(\mathbf{n}) \neq 0 & \text{for every } \mathbf{n},\ j = 1,2,3. \end{array} \right\} \qquad (7.41)$$

This proposition is proved in [1].

From (7.41), one sees that positive definiteness of $\mathbf{A}(\mathbf{n})$, together with $c > 0$, is *sufficient* for material stability, though not necessary. While positive *semi-definiteness* of $\mathbf{A}(\mathbf{n})$ is necessary for material stability, an example put forward in [1] shows that the first two conditions in (7.41) cannot by themselves be sufficient for stability.

The unstable materials arising in the present monograph will all lose stability by virtue of the failure of positive semi-definiteness of the isothermal acoustic tensor at certain states $\mathbf{F}^0, \theta^0$, the specific heat remaining always positive. If θ^0 is one of these temperatures, the *elastic* material whose elasticity tensor is $\mathbf{C}(\cdot, \theta^0)$ must thus lose strong ellipticity at certain deformation gradients.

7.4 A one-dimensional special case: uniaxial strain

We now specialize the theory of thermoelasticity as described above to the particular setting of uniaxial strain in a material that is homogeneous in the reference configuration. Uniaxial strain is often viewed as the appropriate kinematic setting in which to describe shock waves arising in impact experiments; see, for example, [8].

Throughout this section, we assume that body force and heat supply are both absent, so that $\mathbf{b} = \mathbf{0}, r = 0$.

In uniaxial strain, all particles are displaced parallel to a fixed direction, say the x_1-axis in a rectangular Cartesian frame, and the value of the x_1-component of displacement depends only on x_1 and the time t: $u_1 = u_1(x_1, t), u_2 = u_3 = 0$. Moreover, the temperature depends only on x_1 and t: $\theta = \theta(x_1, t)$. To simplify notation, it is convenient to write x, u, and v instead of x_1, u_1, and v_1, respectively, and to indicate x- and t-derivatives by subscripts. Thus, for example, $u_{1,1}(x_1, t) = u_x(x, t)$ and $v_1(x_1, t) = v(x, t) = u_t(x, t)$.

In uniaxial strain, the deformation gradient matrix has the form

$$\underline{F} = \begin{pmatrix} 1+\gamma & 0 & 0 \\ 0 & 1 & 0 \\ 0 & 0 & 1 \end{pmatrix}, \tag{7.42}$$

where $\gamma = \gamma(x, t) = u_x(x, t)$. We write

$$\hat{\Psi}(\gamma, \theta) = \hat{\psi}(\underline{F}, \theta) \tag{7.43}$$

when $\underline{F}$ has the form (7.42). It then follows from the constitutive law (7.17) that

$$\sigma = \hat{\sigma}(\gamma, \theta) = \rho\hat{\Psi}_\gamma(\gamma, \theta), \qquad \eta = \hat{\eta}(\gamma, \theta) = -\hat{\Psi}_\theta(\gamma, \theta) \tag{7.44}$$

where we have written $\sigma = \sigma_{11}$ for the component of nominal stress of primary interest. The alternative version (7.6) of the constitutive law allows us to equivalently write

$$\sigma = \rho\hat{\epsilon}_\gamma(\gamma, \eta), \qquad \theta = \hat{\epsilon}_\eta(\gamma, \eta), \tag{7.45}$$

in terms of the potential $\hat{\epsilon}(\gamma, \eta) = \epsilon(\underline{F}, \eta)$ of specific internal energy expressed as a function of strain and specific entropy.

Since all physical quantities are independent of x_2 and x_3, the momentum equation (7.21) for $j = 1$ then reduces in the absence of body force to

$$(\hat{\sigma}(\gamma, \theta))_x - \rho v_t = 0, \tag{7.46}$$

where $\hat{\sigma}$ is the thermoelastic stress response function for uniaxial strain introduced in (7.44).

Setting $m(\gamma, \theta) = m_{11}(\underline{F}, \theta)$ for the relevant component of the stress–temperature tensor, we may write (7.46) in the expanded form

$$\hat{\sigma}_\gamma(\gamma, \theta)\gamma_x + m(\gamma, \theta)\theta_x - \rho v_t = 0, \tag{7.47}$$

where $\hat{\sigma}_\gamma$ stands for $\partial\hat{\sigma}/\partial\gamma$.

Where γ and v are smooth, their relation to u requires that they satisfy the compatibility equation

$$v_x - \gamma_t = 0. \tag{7.48}$$

For $(i, j) \neq (1, 1)$, the components σ_{ij} of the nominal stress tensor cannot be expressed in terms of $\hat{\Psi}(\gamma, \theta)$ or its derivatives. For $i = 2, 3$, the differential equations

(7.21) for linear momentum, in the absence of body force and with $v_1 = v_2 = 0$, reduce to $\sigma_{21,1} = 0$ and $\sigma_{31,1} = 0$; unless these equations are satisfied, the thermoelastic material under study does not sustain nontrivial states of uniaxial strain parallel to the x_1-axis. To avoid this difficulty, we shall assume that the symmetry of the material is such that, in uniaxial strain, $\sigma_{21} = \psi_{F_{21}}(\underline{F}, \theta) = 0$ and $\sigma_{31} = \psi_{F_{31}}(\underline{F}, \theta) = 0$ when $\underline{F}$ has the form (7.42), so that (7.21) is trivially satisfied for $i=2, 3$. This is the case for isotropic materials, and for certain other material symmetries if the direction of the x_1-axis is suitably oriented. With $\sigma_{21} = \sigma_{31} = 0$, balance of angular momentum implies that $\sigma_{12} = \sigma_{13} = 0$ as well, and that $\sigma_{23} = \sigma_{32}$. Thus in uniaxial strain parallel to the x_1-axis, the nominal stress matrix $\underline{\sigma}$ has the form

$$\underline{\sigma} = \begin{pmatrix} \sigma & 0 & 0 \\ 0 & \sigma_{22} & \sigma_{23} \\ 0 & \sigma_{23} & \sigma_{33} \end{pmatrix}. \tag{7.49}$$

From (6.14) and (7.49), we find that the true stress matrix is then given by

$$\underline{\tau} = \frac{1}{1+\gamma} \begin{pmatrix} (1+\gamma)\sigma & 0 & 0 \\ 0 & \sigma_{22} & \sigma_{23} \\ 0 & \sigma_{23} & \sigma_{33} \end{pmatrix}. \tag{7.50}$$

We set $q = q_{11}$, $K(\gamma, \theta) = K_{11}(\underline{F}, \theta)$ and infer from the heat conduction law (7.18) that

$$q = K(\gamma, \theta)\theta_x. \tag{7.51}$$

Since the x_2- and x_3-components of the heat conduction vector $\mathbf{q}$ depend only on $x_1 = x$, the energy equation (7.22) takes the following form for uniaxial strain in the absence of heat supply:

$$[K(\gamma, \theta)\theta_x]_x + \theta m(\gamma, \theta)v_x - \rho c(\gamma, \theta)\,\theta_t = 0. \tag{7.52}$$

Equations (7.47), (7.48), and (7.52) are the three basic scalar partial differential equations for the three unknowns γ, v, and θ associated with uniaxial strain of a thermoelastic material.

Finally we specialize the respective jump conditions $(7.24)_2$, (7.26), and (7.28) to the case of uniaxial strain:

$$[\![\gamma]\!]\, V + [\![v]\!] = 0, \tag{7.53}$$

$$[\![\sigma]\!] + \rho[\![v]\!]\, V = 0, \tag{7.54}$$

$$(\rho[\![\epsilon]\!] - \langle\sigma\rangle\,[\![\gamma]\!])\, V + [\![q]\!] = 0, \tag{7.55}$$

where we have written $V = V_n$ for the speed of the propagating discontinuity.

REFERENCES

[1] R. Abeyaratne and J.K. Knowles, On the stability of thermoelastic materials. *Journal of Elasticity*, **53** (1999), pp. 199–213.

[2] M.E. Gurtin, The linear theory of elasticity, in *Handbuch der Physik*, edited by S. Flügge, Volume VIa/2. Springer-Verlag, Berlin, 1972.

[3] M.E. Gurtin, *An Introduction to Continuum Mechanics*. Academic Press, New York, 1981.

[4] J. Hadamard, *Leçons sur la propagation des ondes et les équations de la hydro-dynamique*. Hermann et Cie, Paris, 1903.

[5] R.D. James, Finite deformation by mechanical twinning. *Archive for Rational Mechanics and Analysis*, **77** (1981), pp. 143–76.

[6] R.D. James, Phase transformations and non-elliptic free energy functions, in *New Perspectives in Thermodynamics*, edited by J. Serrin. Springer-Verlag, Berlin, 1986, pp. 223–39.

[7] J.K. Knowles and E. Sternberg, On the failure of ellipticity and the emergence of discontinuous deformation gradients in plane finite elastostatics. *Journal of Elasticity*, **8** (1978), pp. 329–79.

[8] Z. Rosenberg, N.K. Bourne, and J.C.F. Millett, Direct measurements of strain in shock-loaded glass specimens. *Journal of Applied Physics*, **79** (1996), pp. 3971–4.

[9] C. Truesdell and W. Noll, The non-linear field theories of mechanics, in *Handbuch der Physik*, edited by S. Flügge, Volume III/3. Springer, Berlin, 1965.

8 Kinetics and Nucleation

8.1 Introduction

In Chapters 3 and 4 we used particular one-dimensional initial–boundary value problems to demonstrate that, because of a massive lack of uniqueness that exists otherwise, elasticity theory must be supplemented with a kinetic law and nucleation condition if it is to be used to model the emergence and evolution of multiphase configurations. As shown there, not only is there a *need* for such information, there is also *room* for it in the theory. A second motivation for a kinetic law, also presented in Chapter 3, arose by casting the quasistatic problem considered there in the framework of standard internal-variable theory; the evolution law characterizing the rate of change of the internal variable in that theory is then the kinetic law.

In the present chapter we provide a third approach to the notion of kinetics, this one from a thermodynamic point of view. In addition to providing a motivation, the discussion here allows us to describe the kinetic law within a general three-dimensional thermoelastic setting.

In Section 8.2 we present the thermodynamic formalism of irreversible processes in a thermoelastic body. Based on this, we introduce the notion of a thermodynamic driving force and the flux conjugate to it, and the notion of a kinetic relation then follows naturally. In Section 8.3 we present some phenomenological examples of kinetic relations, while in Section 8.4 we describe examples of kinetic relations based on various underlying transformation mechanisms. Some remarks on the nucleation condition are made in Section 8.5.

8.2 Thermodynamic fluxes, forces, and kinetics in nonequilibrium processes

In order to illuminate the general idea, we start with a special case. Consider first a thermoelastic body occupying a region $\mathcal{R}$ in a reference configuration undergoing a *smooth* thermomechanical process on a time interval $[t_0, t_1]$. We know from the discussion in Chapter 7 that since the material is thermoelastic, entropy production in the bulk is due solely to heat conduction; further, since the process is smooth, there are no interfaces in the body which might produce any additional entropy. Thus recalling (6.16), (6.19), (6.21), and (7.8), the total rate of entropy production associated with a subregion $\mathcal{D}$ of $\mathcal{R}$ is

$$\Gamma_{\mathcal{D}}(t) = \int_{\mathcal{D}} \mathbf{q} \cdot \frac{\nabla \theta}{\theta^2} \, dV \geq 0. \tag{8.1}$$

Since (8.1) holds for all $\mathcal{D} \subset \mathcal{R}$, and since $\theta > 0$, necessarily

$$\mathbf{q}(\mathbf{x}, t) \cdot \nabla \theta(\mathbf{x}, t) \geq 0 \quad \text{for} \quad \mathbf{x} \in \mathcal{R}, \quad t \in [t_0, t_1]. \tag{8.2}$$

This shows that the heat flux vector $\mathbf{q}$ and the temperature gradient vector $\nabla \theta$ cannot point in opposite directions. Thus one says that "the temperature gradient $\nabla \theta$ drives the heat flux $\mathbf{q}$" and that "$\nabla \theta$ is the *driving force* conjugate to $\mathbf{q}$."

If this body is in thermodynamic equilibrium at an instant t_*, then $\Gamma_{\mathcal{D}}(t_*) = 0$ for all subregions $\mathcal{D} \subset \mathcal{R}$. Therefore

$$\mathbf{q}(\mathbf{x}, t_*) \cdot \nabla \theta(\mathbf{x}, t_*) = 0 \quad \text{for} \quad \mathbf{x} \in \mathcal{R}. \tag{8.3}$$

There is nothing in the description of a thermoelastic material, or the usual principles of continuum mechanics, that provides any information about the driving force $\nabla \theta$ and the flux $\mathbf{q}$ when the body is *not* in thermodynamic equilibrium, and the notion of a kinetic law is intimately related to nonequilibrium thermodynamic processes. According to the theory of irreversible processes (e.g. Sections 14-2 and 14-3 of Callen [12] or Section 14.3.1 of Kestin [18]), such information must be provided empirically, by relating the present value of the flux $\mathbf{q}$ to the present (and possibly past) value of the driving force $\nabla \theta$ (and perhaps to other state variables, such as $\mathbf{F}$ and θ, as well). In the present setting, a natural example of this would be a relation of the form

$$\mathbf{q}(\mathbf{x}, t) = \hat{\mathbf{q}}(\nabla \theta(\mathbf{x}, t), \mathbf{F}(\mathbf{x}, t), \theta(\mathbf{x}, t)) \quad \text{for} \quad \mathbf{x} \in \mathcal{R}, \quad t \in [t_0, t_1], \tag{8.4}$$

where $\hat{\mathbf{q}}$ is a constitutive function in the sense that, in general, it will be different for different materials. The *kinetic law* (8.4) characterizes the response of the body when it is not necessarily in thermodynamic equilibrium; in the present context, it is no more than the *heat conduction law*. In view of the inequality (8.2), the function $\hat{\mathbf{q}}$ must have the property that at each temperature θ and deformation gradient $\mathbf{F}$,

$$\hat{\mathbf{q}}(\mathbf{g}, \mathbf{F}, \theta) \cdot \mathbf{g} \geq 0 \quad \text{for all vectors } \mathbf{g}. \tag{8.5}$$

If $\hat{\mathbf{q}}(\mathbf{g}, \mathbf{F}, \theta)$ is continuous at $\mathbf{g} = \mathbf{0}$, (8.5) shows that $\hat{\mathbf{q}}(\mathbf{0}, \mathbf{F}, \theta) = \mathbf{0}$, that is the heat flux must vanish when the temperature gradient vanishes. This further justifies the term "thermodynamic driving force" for $\nabla \theta$. The continuum theory, on its own, provides no further information about $\hat{\mathbf{q}}$. If the heat flux depends linearly on the temperature gradient, so that $\hat{\mathbf{q}}(\nabla \theta, \mathbf{F}, \theta) = \mathbf{K}(\mathbf{F}, \theta) \nabla \theta$, the 2-tensor $\mathbf{K}$ is the conductivity tensor and (8.5) shows that $\mathbf{K}$ must be positive semidefinite at θ and $\mathbf{F}$: $\mathbf{K}(\mathbf{F}, \theta)\mathbf{g} \cdot \mathbf{g} \geq 0$.

The preceding discussion concerned a smooth process of a thermoelastic body. More generally, if the rate of entropy production $\Gamma_D(t)$ can be expressed as the sum of the products of various thermodynamic driving forces and their conjugate fluxes, then according to the thermodynamic theory of irreversible processes, the evolution of the system is governed by kinetic laws relating the driving forces to the fluxes; for example, see Sections 14-2 and 14-3 of Callen [12].

The particular setting of interest to us involves the propagation of a phase boundary in a thermoelastic body. In this case, we have from (6.16), (6.19), (6.21), and (6.29) that the rate of entropy production associated with the particles in D is

$$\Gamma_D(t) = \int_D \frac{\nabla\theta}{\theta^2} \cdot \mathbf{q} \, dV + \int_{S_t \cap D} \frac{f}{\langle\theta\rangle} V_n \, dA, \qquad (8.6)$$

where the surface of discontinuity S_t corresponds to a phase boundary, f is the driving force on it as introduced in (6.28), and $\langle\theta\rangle$ is the average of the two temperatures on either side of the phase boundary. Thus there are two sources of entropy production in this case: heat conduction and phase transformation. If the body is in thermal equilibrium, $\theta(\mathbf{x}, t) = \text{constant}$, the entropy production rate due to heat conduction vanishes; if the body is in *phase equilibrium*, by which we mean that $f(\mathbf{x}, t)$ vanishes on S_t, the entropy production rate due to the moving phase boundary vanishes. Thus $\nabla\theta/\theta^2$ and $f/\langle\theta\rangle$ are the "agents of entropy production" when a thermoelastic body is not in thermal and phase equilibrium, and the associated fluxes are $\mathbf{q}$, the flux of heat, and ρV_n, the flux of mass across the propagating phase boundary.

The evolution of the various fluxes in a not-necessarily reversible process is governed by *kinetic laws* relating each flux to its conjugate agent of entropy production and possibly other variables. In the present setting, the simplest examples of this would be

Heat conduction:

$$\mathbf{q}(\mathbf{x}, t) = \hat{\mathbf{q}}(\nabla\theta(\mathbf{x}, t), \mathbf{F}(\mathbf{x}, t), \theta(\mathbf{x}, t)) \quad \text{for} \quad \mathbf{x} \in \mathcal{R} - S_t, \quad t \in [t_0, t_1],$$

Phase transition: $\qquad\qquad\qquad\qquad\qquad\qquad\qquad\qquad\qquad\qquad\qquad$ (8.7)

$$V_n(\mathbf{x}, t) = \Phi(f(\mathbf{x}, t), \langle\theta(\mathbf{x}, t)\rangle) \quad \text{for} \quad \mathbf{x} \in S_t, \quad t \in [t_0, t_1].$$

Observe that the heat conduction law holds at all points in the body where the fields are smooth, while the kinetic law for phase transition holds at all points on the phase boundary. Because of the entropy inequality (6.33) and (8.2), the constitutive functions $\hat{\mathbf{q}}$ and Φ must satisfy the requirements

$$\hat{\mathbf{q}}(\mathbf{g}, \mathbf{F}, \theta) \cdot \mathbf{g} \geq 0 \quad \text{for all vectors } \mathbf{g}, \qquad \Phi(f, \theta)f \geq 0 \quad \text{for all scalars } f, \quad (8.8)$$

at each θ and $\mathbf{F}$.

In both quasistatics and dynamics in one dimension, we postulated that the propagation speed of a phase boundary depended on the local state of the material on either side of the interface; see equations (3.43) and (4.22) respectively. A similar

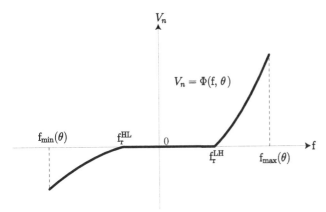

Figure 8.1. Schematic graph of a generic kinetic response function $\Phi(f, \theta)$ as a function of driving force f at fixed temperature θ; f_r^{LH} and f_r^{HL} are the resistances to phase boundary motion and $f_{max}(\theta)$ and $f_{min}(\theta)$ are the maximum and minimum values of driving force. If the nucleation levels of driving force f_n^{LH} and f_n^{HL} were displayed on this figure they would lie outside the open interval (f_r^{HL}, f_r^{LH}). In applying this kinetic law at a phase boundary across which the temperature is discontinuous, the second argument of the function Φ is taken to be the average temperature $\langle\theta\rangle$.

approach in the present setting would suggest the form

$$V_n = \Phi(\mathbf{F}^+, \mathbf{F}^-, \theta^+, \theta^-) \tag{8.9}$$

for the kinetic law that includes $(8.7)_2$ as a special case. In one dimension, we were able to *derive* the form $V_n = \Phi(f)$ from this assumption; in the general theory of the present chapter this is no longer possible so that (8.9) is a true generalization of $(8.7)_2$.

A kinetic relation of the general form $V_n = \Phi(f, \langle\theta\rangle)$ is both homogeneous and isotropic in the sense that the kinetic response of the material is the same at all particles and in all directions. This can be generalized to

$$V_n = \Phi(f, \langle\theta\rangle, \mathbf{n}, \mathbf{x}), \tag{8.10}$$

where the dependence of Φ on $\mathbf{x}$ in S_t reflects inhomogeneity, and the dependence on the unit normal $\mathbf{n}$ on S_t (i.e. the direction of propagation) allows for anisotropy.

8.3 Phenomenological examples of kinetic relations

The continuum theory of thermoelasticity does not provide any information about the kinetic law beyond the inequality $(8.8)_2$. This can only be provided by a more detailed theory that is able to model the microscale processes associated with the transformation of one microstructure (e.g. lattice in the case of a crystalline solid) to another. The kinetic response function Φ might be expected to have the generic characteristics shown schematically in Figure 8.1. As in the purely mechanical theory of Section 3.5, the threshold value $f_r^{LH}(\theta) \geq 0$ of driving force represents the resistance to phase boundary motion, which must be overcome at the temperature

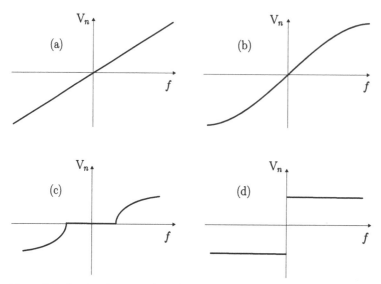

Figure 8.2. Graphs of propagation velocity V_n versus driving force f at fixed temperature θ for various kinetic relations $V_n = \Phi(f, \theta)$: (a) linear, (b) monotonic (c) lattice trapping, and (d) maximally dissipative.

θ in order to cause growth of the high-strain phase; similarly, $f_r^{HL}(\theta) \leq 0$ is the resistance to growth of the low-strain phase for the reverse transition.

Figure 8.2 shows graphs of the propagation speed V_n versus the driving force f at a fixed temperature θ, for various phenomenological kinetic relations $V_n = \Phi(f, \theta)$. A linear kinetic response is depicted in Figure 8.2(a). Figure 8.2(b) shows a general monotonically rising kinetic curve describing a situation where the larger the driving force, the larger is the propagation speed. The kinetic curve shown in Figure 8.2(c) describes a circumstance where, if the driving force is sufficiently small, the phase boundary does not propagate; sufficiently large values of driving force are required for the phase transition to progress. This phenomenon is sometimes referred to as "lattice trapping." The kinetic curve shown in Figure 8.2(d) is called maximally dissipative (in quasistatics) for reasons entirely analogous to those that lead to the dynamically maximally dissipative kinetic law of Figure 4.7. If a phase boundary propagates without dissipation, then the driving force must vanish and one has the "Maxwell kinetic law" $f = 0$. Finally, nonmonotonic kinetic relations are also of interest in describing phase boundary motion in which the interface alternately stops and propagates, exhibiting "stick-slip" behavior.

8.4 Micromechanically based examples of kinetic relations

The role of the kinetic law is to provide the continuum theory with an appropriately homogenized characterization of the dynamics associated with the microscale process of transformation. In contrast to the phenomenological models described above, a physics-based kinetic relation has to be derived, or at least motivated, by modeling and analyzing the microscale processes. In order to provide a glimpse of some of the sorts of modeling that one might consider, we sketch out five examples here. It should be emphasized that our goal is to illustrate some different sorts of models rather than to develop any physically accurate models.

In each of the calculations below we assume that the motion occurs isothermally at a fixed temperature. Therefore we shall not display dependencies on θ explicitly (except in the model of thermally activated processes). The kinetic response functions $\Phi(f, \theta)$ determined below are thus valid at the fixed temperature underlying the calculations.

The simplest form of kinetic relation governing a three-dimensional motion of a phase boundary relates the driving force on it to its normal velocity of propagation: $V_n = \Phi(f)$. Perhaps it is worth remarking that since the kinetic response function Φ is a function of a single scalar independent variable, one set of experiments, say uniaxial tension tests, completely determines Φ; and the function Φ thus determined characterizes all motions of this interface (at that temperature) even if the conditions are not those of uniaxial stress. If the deformation field is inhomogeneous, and the phase boundary is curved, one would use this kinetic relation locally, at each point along the interface, relating the driving force at that point to the normal velocity of propagation of that point. Consequently, in order to determine the kinetic response function it is sufficient to study, say, one-dimensional motions of the body. All of the kinetic relations derived below are developed on the basis of convenient one-dimensional models, but for the reasons just mentioned, they are applicable more generally.

In order to keep the analysis as simple as possible, the first three examples will be developed within the following particularly simple context: consider a one-dimensional bar characterized by the biquadratic strain energy density

$$
W(\gamma) = \begin{cases} \dfrac{\mu}{2}\gamma^2, & \gamma \leq \gamma_*, \\[2mm] \dfrac{\mu}{2}(\gamma - \gamma_T)^2, & \gamma \geq \gamma_*, \end{cases} \tag{8.11}
$$

with the associated bilinear stress response function

$$
\hat{\sigma}(\gamma) = W'(\gamma) = \begin{cases} \mu\gamma, & \gamma < \gamma_*, \\[2mm] \mu(\gamma - \gamma_T), & \gamma > \gamma_*; \end{cases} \tag{8.12}
$$

see Figure 8.3. Here W is related to the Helmholtz free energy function ψ through $W(\gamma) = \rho\psi(\gamma)$. Let

$$
\gamma_* = \frac{\gamma_T}{2}, \qquad \sigma_* = \mu\gamma_*, \qquad f_* = \frac{\mu\gamma_T^2}{2}. \tag{8.13}
$$

Two phase equilibrium states are possible when the stress lies in the range $-\sigma_* < \sigma < \sigma_*$; the strains on the two sides of a phase boundary in mechanical equilibrium are

$$
\gamma^+ = \frac{\sigma}{\mu}, \qquad \gamma^- = \frac{\sigma}{\mu} + \gamma_T, \tag{8.14}
$$

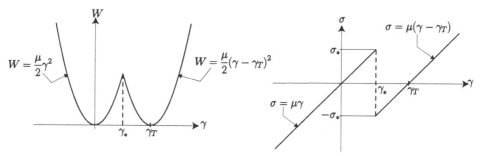

Figure 8.3. Biquadratic energy and bilinear stress–strain relation.

when the high-strain phase is on the left and the low-strain phase is on the right. The driving force on such a phase boundary is

$$f = W(\gamma^+) - W(\gamma^-) - \sigma(\gamma^+ - \gamma^-) = \sigma\gamma_T. \qquad (8.15)$$

The potential energy function is

$$P(\gamma, \sigma) = W(\gamma) - \sigma\gamma. \qquad (8.16)$$

8.4.1 Viscosity–strain gradient model. In this model we augment the elastic constitutive relation $\sigma = \hat{\sigma}(\gamma)$ by adding to it a linear viscosity term characterized by a coefficient ν and a linear strain-gradient term characterized by a parameter λ. The viscosity allows for dissipation and the strain gradient allows the sharp phase boundary to be spread over a narrow transition zone. Thus the constitutive relation for stress, when augmented with these additional effects, reads

$$\sigma = \hat{\sigma}(\gamma) + \rho\nu\gamma_t - \rho\lambda\gamma_{xx}, \quad \nu > 0, \quad \lambda > 0, \qquad (8.17)$$

where $\hat{\sigma}$ continues to be given by (8.12). We consider a quasistatic motion and so the stress is spatially uniform in the bar. We wish to model a phase boundary of the purely elastic theory, which is a sharp interface that travels at a speed V_n and separates the strains $\gamma^\pm$. In the augmented theory, this is described by a traveling wave in an infinite bar, propagating at the speed V_n and approaching the strains $\gamma^\pm$ in the far field:

$$\gamma = \gamma(\xi) \quad \text{for} \ -\infty < \xi < \infty, \qquad \gamma(\xi) \to \gamma^\pm \quad \text{as} \ \ \xi \to \pm\infty, \quad (8.18)$$

where

$$\xi = x - V_n t. \qquad (8.19)$$

Without loss of generality, the respective regions $\xi > 0$ and $\xi < 0$ may be associated with the low- and high-strain phases of the material. Substituting (8.18)$_1$ into (8.17) leads to the differential equation

$$\rho\lambda\gamma''(\xi) + \rho\nu V_n\gamma'(\xi) + \sigma - \hat{\sigma}(\gamma(\xi)) = 0, \quad -\infty < \xi < \infty, \qquad (8.20)$$

where σ is the uniform stress in the bar. The traveling wave $\gamma(\xi)$ must satisfy this equation and the remote boundary conditions (8.18)$_2$. In light of the smoothness characteristics of the stress response function (8.12), and the order of the differential equation (8.20), the strain field $\gamma(\xi)$ must be continuous and have a continuous first derivative. It is convenient to introduce the nondimensional propagation speed v and the material parameters c and ω defined by

$$v = \frac{V_n}{\omega c}, \qquad c = \sqrt{\frac{\mu}{\rho}}, \qquad \omega = 2\frac{\sqrt{\lambda}}{v}. \qquad (8.21)$$

Explicitly solving the differential equation (8.20) by using (8.12), and enforcing the boundary conditions (8.18)$_2$ and the continuity of $\gamma(\xi)$ at $\xi = 0$, shows that the strain field is given by

$$\gamma(\xi) = \begin{cases} \gamma^+ + (\gamma_* - \gamma^+)e^{p_2\xi}, & \xi \geq 0, \\ \gamma^- - (\gamma^- - \gamma_*)e^{p_1\xi}, & \xi \leq 0, \end{cases} \quad \text{where} \quad \begin{matrix} p_1 \\ p_2 \end{matrix} = \frac{2c}{\omega}\left(-v \pm \sqrt{v^2 + 1}\right).$$
$$(8.22)$$

Requiring $\gamma'(\xi)$ to be continuous at $\xi = 0$ then leads to the following relation between the driving force f and the propagation speed v:

$$f = \phi(v) = f_* \frac{v}{\sqrt{1 + v^2}} = f_* \frac{V_n}{\sqrt{\omega^2 c^2 + V_n^2}}; \qquad (8.23)$$

here f and f_* were given in (8.15) and (8.13)$_3$ respectively. Equation (8.23) characterizes the kinetic law for a quasistatic transition controlled by viscosity and strain gradient effects.

Since $\omega \to \infty$ when $v \to 0$ with $\lambda > 0$ fixed, one sees from (8.23)$_2$ that $f = 0$ in the limit of vanishing viscosity. This also follows directly from an energy balance. Thus if $v = 0$, phase boundary motion in the sharp-interface limit as $\lambda \to 0$ is dissipation free.

Since $\omega \to 0$ when $\lambda \to 0$ with $v > 0$ fixed, (8.23)$_2$ shows that the limiting kinetic relation in the sharp-interface theory is $f = f_*$, which is maximally dissipative in the sense of Section 8.3.

Further details concerning models of this sort, including models that include inertial and/or thermal effects, can be found, for example, in [2, 27–31, 32–35]. For an extensive mathematical discussion of kinetic relations arising from traveling wave solutions of regularized hyperbolic systems of conservation laws, the reader may consult LeFloch [21].

8.4.2 Thermal activation model. In order to construct this next model, consider a quasistatic motion of the elastic bar and consider an instant at which the stress has the value σ. Suppose there is a phase boundary located at $x = s$, and the phase boundary is propagating into the low-strain phase at a speed $V_n = \dot{s}$. The strains $\gamma^\pm$ on either side of the phase boundary are given by (8.14). The potential energy function $P(\cdot, \sigma)$ given by (8.16) has local minima at γ^- and γ^+ and a local maximum between them at γ_*; see Figure 8.4. As the phase boundary moves through the bar, particles immediately ahead of it "jump" from the higher energy

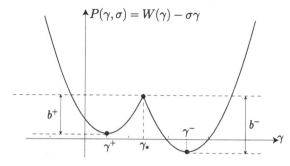

$$P(\gamma, \sigma) = W(\gamma) - \sigma\gamma$$

Figure 8.4. Graph of potential energy P versus strain γ at a fixed stress σ for the biquadratic model (8.11) and (8.16). The energy barriers are denoted by $b^{\pm}$. The figure has been drawn for $\sigma > 0$ in which case the high-strain energy well is lower than the low-strain energy well. The locations $\gamma^{\pm}$ of the local minima are given by (8.14).

well of P to the lower one, and an associated kinetic relation can be constructed by viewing this process of jumping at the atomic scale. It should be noted that at the atomic scale, some atoms from *each* energy well jump into the other energy well, and it is the net effect of this process that is reflected at the macroscopic scale. The atoms behind the phase boundary jump from the energy well at γ^- to that at γ^+ and have to cross over an energy barrier

$$b^- = P(\gamma_*, \sigma) - P(\gamma^-, \sigma) = \frac{(\sigma_* - \sigma)^2}{2\mu} + \sigma\gamma_T \qquad (8.24)$$

in order to do so; see Figure 8.4. Similarly the atoms ahead of the phase boundary jump from the energy well at γ^+ to that at γ^- and in doing so must cross over an energy barrier

$$b^+ = P(\gamma_*, \sigma) - P(\gamma^+, \sigma) = \frac{(\sigma_* - \sigma)^2}{2\mu}. \qquad (8.25)$$

In a "thermally activated process," the probability that the energy of an atom is at least as great as b is $\exp(-b/k\theta)$ where k is Boltzmann's constant and θ is the absolute temperature. Moreover, the average rate at which atoms jump from one minimum to the other in such a process is proportional to the probability of exceeding the corresponding energy barrier. Thus in a thermally activated transition, the velocity of the phase boundary, which is a macroscopic measure of the net rate at which atoms change from the low-strain phase to the high-strain phase, is the difference between the rates of the two atomic transitions and is therefore given by

$$V_n = R\left\{\exp\left(-\frac{b^+}{rk\theta}\right) - \exp\left(-\frac{b^-}{rk\theta}\right)\right\}. \qquad (8.26)$$

Here R is related to the frequency with which an atom attempts to cross over the relevant energy barrier and the factor r converts the units of P (and therefore b) to those of energy per atom. Substituting for the various quantities in this expression by using (8.24), (8.25), and (8.15) leads to the following explicit kinetic relation between the driving force f and the propagation speed V_n for a thermally activated transition:

$$V_n = \Phi(f) = 2R\ \exp\left(-\frac{f^2 + f_*^2}{4rk\theta f_*}\right)\sinh\left(-\frac{f}{2rk\theta}\right). \qquad (8.27)$$

For a more complete description of models of this sort, see, for example, [3, 6, 7, 15, 16, 22, 23].

8.4.3 Propagation through a row of imperfections.

Consider a two-phase equilibrium state of the bar at a stress σ involving a phase boundary at $x = s$. Suppose for convenience that the bar has unit length and unit cross-sectional area, so that s represents the volume fraction of the high-strain phase. It is readily shown from (8.11) and (8.14) that the total elastic strain energy in the bar is

$$E_e(\sigma) = \frac{\sigma^2}{2\mu}. \tag{8.28}$$

Suppose further that the surface energy associated with the phase boundary must be accounted for and that it can be characterized by

$$E_s = E_s(s); \tag{8.29}$$

a physical example in which such a model is applicable is described in Chapter 14. Next, if the bar is subjected to a dead-loading in which the stress σ is prescribed, then the energy of the loading device is

$$E_\ell(s, \sigma) = -\sigma\delta = -\frac{\sigma^2}{\mu} - \sigma s\gamma_T, \tag{8.30}$$

where $\delta = s\gamma^- + (1-s)\gamma^+ = \sigma/\mu + s\gamma_T$ is the elongation of the bar.

Finally, suppose that there are a large number of small imperfections along the length of the bar, equally spaced at a distance $\varepsilon/(2\pi)$ apart. Suppose further that each imperfection causes a slight increase in the energy of the system when the phase boundary passes across it. This suggests that the energy associated with the imperfections can be modeled by a series of energy wells, a distance $\varepsilon/(2\pi)$ apart, with the location of the local maxima between energy wells corresponding to the locations of the imperfections. A simple model for this energy is then

$$E_i(s, \varepsilon) = a\varepsilon \cos(s/\varepsilon), \tag{8.31}$$

where a and $\varepsilon \ll 1$ are constants. The smallness of the amplitude $a\varepsilon$ in (8.31) corresponds to the case where the imperfections are small, that is, they cause only a slight rise in energy.

The total energy E_ε of the system consisting of the bar and loading device is given by

$$E_\varepsilon(s, \sigma) = E_e(\sigma) + E_s(s) + E_\ell(s, \sigma) + E_i(s, \varepsilon). \tag{8.32}$$

The problem characterized in this way falls into an internal-variable framework: the stress σ and the volume fraction s are, respectively, the macroscopic variable and the internal variable in the potential $E_\varepsilon(s, \sigma)$. Specifically, the driving force conjugate to s is $-\partial E_\varepsilon/\partial s$. Suppose that for slow motions of the phase boundary one can adequately model its kinetics by a linear relation between the driving force and $\dot{s}$, the rate of change of the internal variable:

$$V_n = -\mathsf{M}\frac{\partial E_\varepsilon}{\partial s}, \tag{8.33}$$

where the proportionality constant M is called the mobility and we have set $V_n = \dot{s}$. Given a quasistatic loading history $\sigma = \sigma(t)$, the evolution of the volume fraction $s(t)$ is determined by solving the ordinary differential equation (8.33), which is of the form $\dot{s}(t) = g(s(t), \sigma(t), \varepsilon)$, subject to an appropriate initial condition $s(0) = s_o$. For a large number of small imperfections one would take the limit $\varepsilon \to 0$ of this kinetic relation. Abeyaratne, Chu, and James [1] have shown that the solution $s_\varepsilon(t)$ of (8.33) converges uniformly in this limit, $s_\varepsilon(t) \to s(t)$ as $\varepsilon \to 0$, where $s(t)$ satisfies the effective kinetic relation

$$V_n = \Phi(f) = \mathsf{M} \begin{cases} (f^2 - a^2)^{1/2} & \text{for} \quad f \geq a, \\ 0 & \text{for} \quad -a \leq f \leq a, \\ -(f^2 - a^2)^{1/2} & \text{for} \quad f \leq -a; \end{cases} \tag{8.34}$$

see also Chapter 14 and Bhattacharya [11]. Here f is the driving force *in the absence of the imperfections*,

$$f = -\frac{\partial}{\partial s}(E_e + E_s + E_\ell) = \sigma \gamma_T - E_s'(s), \tag{8.35}$$

and a is the parameter introduced in (8.31), which is a measure of the energy associated with the imperfections. Note that according to the kinetic relation (8.34), the volume fraction does not change at small values of the driving force and grows only when it exceeds a critical value determined by the imperfection strength. A graph of the kinetic law has the qualitative features shown in Figure 8.2(c).

8.4.4 Kinetics from atomistic considerations.

In principle, the kinetics of a phase-transforming crystal are derivable from a knowledge of the interaction potentials among the atoms of the underlying lattice. A simple analysis along these lines has been undertaken by Purohit [25], who considered the dynamics of a one-dimensional chain of N identical atoms. Each atom is assumed to interact with its nearest and next-nearest neighbors; the potentials for these two interactions are assumed to have two wells and are such that the associated two-phase force–stretch relation between atoms is trilinear, with the same stiffness in each phase. The atom at one end of the chain is subjected to a given velocity v_0, while the atom at the opposite end is fixed, thus modeling tensile impact of the chain. The resulting classical dynamical system consists of a dissipation-free set of N second-order nonlinear ordinary differential equations, which is solved under given initial conditions that at first are taken to be those for atoms at rest in their "low-strain" wells, corresponding to a chain at zero absolute temperature. For $N = 300$, the initial-value problem is solved numerically for various choices of the imposed impact velocity v_0. For sufficiently small v_0, the solution corresponds to a small-amplitude oscillation of the atoms about their initial states that propagates away from the impacted end of the chain at the "sonic" wave speed for the low-strain phase of the potentials. For larger values of v_0, such a small-amplitude, low-strain-phase sonic precursor is followed by a small oscillation about a mean amplitude much greater than that of the precursor. With "strain" defined as the ratio of the difference in displacements of successive atoms to the initial interatomic spacing, the strain involved in the trailing large-amplitude wave lies in the high-strain phase of the interatomic potentials. The "phase boundary" is the propagating abrupt change in the mean amplitude from precursor to follower. Thus the response of the chain-of-atoms model is analogous to that of the continuum mechanical model discussed earlier in Sections 4.3.1 and 4.3.2 in that there is a weak-impact regime without a phase transition for small v_0 and a two-wave regime with a phase change for stronger impacts[1].

To determine the kinetic relation implied by the dynamics of the chain of atoms, Purohit [25] begins with the continuum-mechanical representation of the driving force f at a phase boundary. For the trilinear material with identical moduli μ in the two phases, this is

$$f = \frac{\mu \gamma_T}{2}(\gamma^+ + \gamma^- - \gamma_M - \gamma_m), \qquad (8.36)$$

where $\gamma^\pm$ are the strains on either side of the phase boundary, γ_T is the transformation strain, and γ_M, γ_m are the strains that delimit the two phases. Making use of the numerically calculated strains $\gamma^\pm$ and phase boundary speed V_n for various values of the impact velocity, Purohit calculates the relation between f and V_n for the 300-atom system. He finds that f increases monotonically with V_n; the numerically calculated points in a V_n, f-plane are described quite accurately by an analytical relation $f = \phi(V_n)$, in which $\phi(V_n)$ is the maximally dissipative kinetic response function for the particular trilinear material he considered.

Purohit also explores the effect of attributing initial velocities to the atoms, corresponding to a chain at nonzero absolute temperature.

Other one-dimensional atomistic calculations have been carried out, for example, by Puglisi and Truskinovsky [26], who study quasistatic processes, and by Krumhansl and Schrieffer [20], Balk, Cherkaev, and Slepyan [9, 10], and Kresse and Truskinovsky [19], who are concerned with dynamics.

A kinetic relation for twin boundary motion has been developed through an analysis based on lattice dynamics by Abeyaratne and Vedantam [5] and applied to a CuAlNi alloy. A lattice-dynamical simulation of a shock-induced phase transition in iron has been carried out by Kadau et al. [17]. Their results exhibit both the two-wave structure for moderate shock strength and the overdriven response for strong shocks predicted by the model described earlier in Section 4; they do not address the issue of a kinetic relation for the transition.

---

[1] In the particular discrete model of [25], there is no counterpart of the "overdriven" response exhibited in the continuum case, see Section 4.3.3, because the linear force–stretch relations between atoms in each phase have the same moduli.

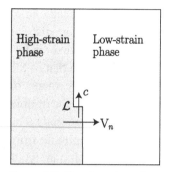 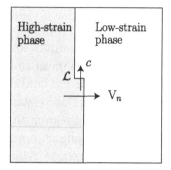

Figure 8.5. The phase boundary propagates to the right with an effective velocity V_n as a result of a ledge, or "step," $\mathcal{L}$, which propagates transversely *along* the phase boundary with speed c. The two figures correspond to two instants of time, showing the ledge $\mathcal{L}$ in two locations.

8.4.5 Frenkel–Kontorowa model. In this model, the motion of a planar phase boundary normal to itself occurs through the mechanism of "ledge" propagation, in which a small step propagates transversely *along* the phase boundary as shown schematically in Figure 8.5. When a ledge $\mathcal{L}$ traverses the entire length of the phase boundary, the phase boundary has advanced by one step. The effective velocity of phase boundary propagation, V_n, is therefore related to the speed of ledge propagation, c, the size of a step, and the dimensions of the phase boundary.

The simplest mathematical model describing this mechanism is as follows: the atoms along the phase boundary are modeled by a one-dimensional chain of particles, each of mass m. In its reference configuration, the nth particle is located at $x_n = nd$, and during a motion, its displacement is denoted by $u_n(t)$. The two-phase nature of the surrounding material is accounted for by associating with each particle a two-well energy function $h(u)$. Each particle is connected to its neighboring particles by a linear spring of stiffness k and natural length d. Finally, the effect of the macroscopic shear stress on the phase boundary is accounted for by a force $\mathcal{F}$ that acts on each particle. The equation of motion of the $n th$ particle is therefore

$$m\ddot{u}_n = \mathcal{F} - h'(u_n) + k(u_{n+1} - 2u_n + u_{n-1}), \quad n = 0, \pm 1, \pm 2, ..., \quad (8.37)$$

Frenkel and Kontorowa [14]. A more detailed discussion of the Frenkel–Kontorowa model and its generalizations can be found, for example, in [4, 5, 8, 24]. A description of how such a model can be used to describe twinning in a three-dimensional crystal can be found in [5].

In order to illustrate the ideas with minimum analytical complication, suppose that the two-well energy potential is biquadratic:

$$h(u) = \begin{cases} \dfrac{\kappa}{2}u^2, & u \leq d/2 \quad \text{(low-strain phase),} \\[2mm] \dfrac{\kappa}{2}(u - d)^2, & u \geq d/2 \quad \text{(high-strain phase),} \end{cases} \quad (8.38)$$

which has energy wells at $u = 0$ and $u = d$.

We wish to model the kinetics of the transformation from the low-strain phase to the high-strain phase. Consider therefore a motion in which all particles are initially in the low-strain energy well and steadily advance, one particle at a time, from that energy well to the high-strain one. Specifically, suppose that the nth particle remains in the low-strain energy well for all times less than some critical instant $n\tau$; at $t = n\tau$ it moves into the high-strain energy well and then remains there for all subsequent times. This motion continues steadily with the $(n + 1)$th particle moving into the high-strain energy well at time $t = (n + 1)\tau$ and so on. Note that at the instant $t = n\tau$, all particles behind the nth one belong to the high-strain phase and therefore have undergone the transformation; all particles ahead of the nth one are still in the original phase, and so there is a "discontinuity" at the nth particle, which is the ledge $\mathcal{L}$. The speed with which this discontinuity propagates is

$$c = d/\tau. \tag{8.39}$$

In this steady motion the displacement of the nth particle at time t equals the displacement of the $(n + m)th$ particle at time $t + m\tau$: $u_n(t) = u_{n+m}(t + m\tau)$, and so the displacement field must admit the traveling wave representation $u_n(t) = u(\xi)$ where $\xi = nd - ct$. Note that $\xi < 0$ is associated with the high-strain phase and $\xi > 0$ is associated with the low-strain phase.

In order to develop a continuum approximation of this problem, let x be a continuous coordinate along the spring-mass chain, and consider a displacement field $u(x, t)$, $-\infty < x < \infty$, $t > 0$, which is such that $u(nd, t) = u_n(t)$. The equation of motion (8.37) can then be written as

$$mu_{tt}(x, t) = \mathcal{F} - h'(u(x, t)) + k[u(x + d, t) - 2u(x, t) + u(x - d, t)],$$
$$x = nd, \ n = 0, \pm1, \pm2, \ldots \tag{8.40}$$

Replacing $u(x \pm d, t)$ by the first five terms in its Taylor expansion, and using the traveling wave representation $u(x, t) = u(\xi)$, $\xi = x - ct$, leads to the ordinary differential equation

$$\frac{kd^4}{12}u''''(\xi) + (kd^2 - mc^2)u''(\xi) + \mathcal{F} - h'(u(\xi)) = 0, \quad -\infty < \xi < \infty. \tag{8.41}$$

In order to determine the boundary conditions to be imposed on (8.41) we note first that in equilibrium, when the displacement field $u(\xi)$ is constant, this constant value is given according to (8.41) by the roots of the equation $h'(u) = \mathcal{F}$. For the biquadratic potential (8.38) this equation has two roots corresponding to the two phases:

$$u^+ = \frac{\mathcal{F}}{\kappa}, \qquad u^- = \frac{\mathcal{F}}{\kappa} + d. \tag{8.42}$$

Suppose that the material far ahead of the ledge is in equilibrium (in the low-strain phase). Then we impose the following boundary conditions:

$$\left.\begin{array}{ll} u(\xi) \to u^+ = \mathcal{F}/\kappa & \text{as } \xi \to +\infty, \\[2mm] u(\xi) \geq d/2 & \text{as } \xi \to -\infty. \end{array}\right\} \tag{8.43}$$

Equation $(8.43)_2$ states that, far behind the interface, $u(\xi)$ must lie in the displacement range corresponding to the high-strain phase. The stricter requirement that the material far behind the ledge be in *equilibrium* in the high-strain phase can be shown to be too stringent.

One can readily solve (8.41) and (8.43) using (8.38) on each of the ranges $\xi < 0$ and $\xi > 0$ separately. Requiring u and its first three derivatives to be continuous at $\xi = 0$, and enforcing the additional continuity condition $u(0) = d/2$, leads to the traveling wave solution

$$u(\xi) = \begin{cases} \left(\dfrac{d}{2} - u^+\right) e^{-b\xi/d} + u^+ & \text{for } \xi > 0, \\[4mm] -\left(\dfrac{d}{2} - u^+\right) e^{b\xi/d} - 2u^+ \cos\dfrac{r\xi}{d} + u^- & \text{for } \xi < 0, \end{cases} \tag{8.44}$$

as well as the following kinetic relation between the force $\mathcal{F}$ and the dimensionless propagation speed v of the ledge:

$$\mathcal{F} = \Phi(v) = \frac{\kappa d}{4}\left\{1 - \frac{1 - v^2\Omega^2}{\sqrt{(1 - v^2\Omega^2)^2 + \Omega^2/3}}\right\}, \tag{8.45}$$

where we have set

$$v = \frac{c}{\sqrt{\kappa d^2/m}}, \qquad \Omega^2 = \frac{\kappa}{k}, \tag{8.46}$$

$$\left.\begin{array}{ll} r = \left[6(1 - v^2\Omega^2) + \sqrt{36(1 - v^2\Omega^2)^2 + 12\Omega^2}\right]^{1/2} & > 0, \\[3mm] b = \left[-6(1 - v^2\Omega^2) + \sqrt{36(1 - v^2\Omega^2)^2 + 12\Omega^2}\right]^{1/2} & > 0. \end{array}\right\} \tag{8.47}$$

Note that for the Frenkel-Kontorowa kinetic law (8.45) there is a minimum force, $\mathcal{F}_{\min}$, required to propagate the ledge. By setting $v = 0$ in (8.45), this value is

$$\mathcal{F}_{\min} = \frac{\kappa d}{4}\left\{1 - \frac{1}{\sqrt{1 + \Omega^2/3}}\right\}. \tag{8.48}$$

Observe from $(8.44)_2$ that the displacement field behind the ledge is oscillatory (and therefore not in equilibrium). These oscillations are reminiscent of those observed by Purohit [25] in his atomistic model; see Section 8.4.4.

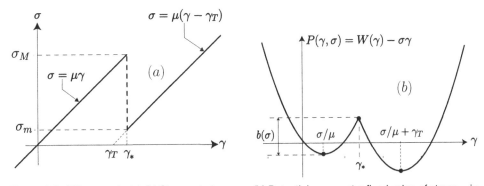

Figure 8.6. Bilinear material. (a) Stress–strain curve. (b) Potential energy at a fixed value of stress σ in the interval $\sigma_0 < \sigma < \sigma_M$, where $\sigma_0 = (\sigma_M + \sigma_m)/2$ is the Maxwell stress.

8.5 Nucleation

A kinetic relation describes the propagation of an *existing* phase boundary. It does not address the issue of when a new phase (or equivalently, a new phase boundary) *first* appears – or "nucleates" – within a configuration that involves a single phase. As demonstrated explicitly in the example problems in Chapters 3 and 4 the phenomenon of nucleation is distinct from propagation and must therefore be described by a criterion other than the kinetic relation. We therefore require a "nucleation criterion."

A complete discussion and analysis of nucleation are beyond the scope of this monograph, not to mention the abilities of its authors! A good description of the various physical effects that must be accounted for, as well as the underlying processes, can be found in Chapter 10 of Christian [13], especially Section 52.

Here we wish to describe a nucleation criterion that elaborates and generalizes the criterion used in Sections 3.5 and 4.4. We present the ideas in the simplest context: that of the purely mechanical theory of equilibria in a one-dimensional bar composed of a two-phase material described in Chapter 3. Thermal effects are not taken into account, but the effect of surface energy at the phase boundary, omitted from the discussion in Chapter 3, is now included and plays a significant role. According to the criterion to be developed here, nucleation results in a portion of the bar of nonzero length switching instantaneously from one phase to the other. The length of this "nucleus" vanishes in the limit of zero surface energy, reducing the theory sketched in the present section to one that is consistent with the model used in the earlier chapters.

Consider a bar that occupies the interval $0 \le x \le L$ in a reference configuration. Suppose the left end is fixed, and a given displacement δ is prescribed at the right end. The material of the bar is assumed to have a bilinear, two-phase stress–strain curve[2] with modulus μ common to both phases, as shown in Figure 8.6(a); the

---

[2] This is a special case of the trilinear material of Figure 2.7 for which $\gamma_M = \gamma_m$ and the moduli of the high- and low-strain phases coincide.

figure identifies special material parameters $\gamma_*, \sigma_M, \sigma_m$, and γ_T that will arise in the discussion to follow.

The potential energy function for this material is

$$P(\gamma, \sigma) = \begin{cases} \frac{1}{2}\mu\gamma^2 - \sigma\gamma, & -1 < \gamma < \gamma_*, -\infty < \sigma < \infty, \\ \frac{1}{2}\mu(\gamma - \gamma_T)^2 + \sigma_0\gamma_T - \sigma\gamma, & \gamma \geq \gamma_*, \quad -\infty < \sigma < \infty, \end{cases} \tag{8.49}$$

where $\sigma_0 = (\sigma_M + \sigma_m)/2 = \mu\gamma_* - \mu\gamma_T/2$ is the Maxwell stress. Figure 8.6(b) shows a graph of P versus γ at a fixed σ greater than the Maxwell stress.

First consider a homogeneous equilibrium configuration in which the bar is entirely in the low-strain phase for the given δ. The corresponding uniform strain and stress fields are

$$\gamma(x) = \delta/L, \qquad \sigma(x) = \mu\delta/L, \quad 0 \leq x \leq L. \tag{8.50}$$

Next consider a different deformation with the same δ wherein the bar is in a two-phase equilibrium state with the segment $0 < x < s$ being in the high-strain phase. The corresponding piecewise uniform strain and stress fields are easily shown to be

$$\gamma(x) = \begin{cases} \delta/L + \gamma_T(1 - s/L), & 0 \leq x < s, \\ \delta/L - \gamma_T s/L, & s < x \leq L, \end{cases}$$

$$\sigma(x) = \mu(\delta/L - \gamma_T s/L), \quad 0 \leq x \leq L. \tag{8.51}$$

If a pure low-strain phase configuration and a two-phase configuration are to exist at the same elongation δ, it is necessary that

$$\gamma_* - \gamma_T < \delta/L < \gamma_*, \tag{8.52}$$

as we shall assume.

Here we view the phenomenon of nucleation as follows: if the elongation δ is slowly increased from zero, nucleation will occur at a critical value $\delta = \delta_{\text{nuc}}$, with the result that the bar switches instantaneously from a uniform low-strain phase equilibrium state to a two-phase equilibrium state[3]. We suppose that the particles for which $0 \leq x \leq s_{\text{nuc}}$ have jumped to the high-strain phase at nucleation, so that s_{nuc} is the size of the nucleus. A proper nucleation criterion will determine δ_{nuc} and s_{nuc} in terms of properties of the bar[4].

The prescribed elongation is assumed to increase continuously, so that δ has the same value just before ("pre") and just after ("post") nucleation:

$$\delta_{\text{pre}} = \delta_{\text{post}} = \delta_{\text{nuc}}; \tag{8.53}$$

---

[3] We model nucleation as occurring instantaneously. In some circumstances, nucleation must be viewed as a time-dependent process; see, for example, Section 50 of Christian [13].
[4] If the elongation continues to slowly increase after nucleation, the phase boundary created at $x = s_{\text{nuc}}$ will propagate according to the kinetic relation.

δ_{nuc}/L is permitted to lie anywhere in the interval (8.52). On the other hand, the values of stress just before and just after nucleation are different as seen from (8.50) and (8.51):

$$\sigma_{pre} = \mu\delta_{nuc}/L, \qquad \sigma_{post} = \sigma_{pre} - \mu\gamma_T\, s_{nuc}/L. \qquad (8.54)$$

Thus in this displacement-controlled process, the stress decreases discontinuously after nucleation. For a stress-controlled process, one can show that the elongation *increases* discontinuously upon nucleation to accommodate the newly created nucleus.

To begin with, we permit the pre-nucleation stress σ_{pre} to lie anywhere[5] in the interval $\sigma_m \leq \sigma \leq \sigma_M$.

Consider a generic particle that changes phase upon nucleation; in the reference configuration, such a particle lies in the segment $0 < x < s_{nuc}$ of the bar. The strains at this particle just before and just after nucleation are given by

$$\gamma_{pre} = \delta_{nuc}/L, \qquad \gamma_{post} = \gamma_{pre} + \gamma_T(1 - s_{nuc}/L). \qquad (8.55)$$

From (8.49), (8.54), and (8.55), one can compute the values of the potential energy of the generic particle just before and just after nucleation:

$$\left. \begin{array}{l} P_{pre} = P(\gamma_{pre}, \sigma_{pre}) = -\sigma_{pre}^2/(2\mu), \\[2mm] P_{post} = P(\gamma_{post}, \sigma_{post}) = -\sigma_{post}^2/(2\mu) - \gamma_T(\sigma_{post} - \sigma_0). \end{array} \right\} \qquad (8.56)$$

Assumption 1. *For nucleation to occur at a generic particle, it is necessary that*

$$P_{post} < P_{pre}. \qquad (8.57)$$

From (8.54) and (8.56), it follows that (8.57) is equivalent to

$$\sigma_{pre} > \frac{\sigma_0 + \mu\,\gamma_T\left[s_{nuc}/L - \frac{1}{2}(s_{nuc}/L)^2\right]}{1 - (s_{nuc}/L)}. \qquad (8.58)$$

One can show that there is an interval of values of s_{nuc} of the form $0 < s_{nuc} < \bar{s}_{nuc}$ on which (8.58) holds if and only if $\sigma_{pre} > \sigma_0$. Said differently, there is an available nucleus of particles, each of whose potential energies would decrease upon nucleation as required by Assumption 1, if and only if $\sigma_{pre} > \sigma_0$. Thus for nucleation to occur, Assumption 1 has allowed us to conclude that

$$\sigma_0 < \sigma_{pre} \leq \sigma_M. \qquad (8.59)$$

As a function of strain at a fixed stress whose value exceeds the Maxwell stress, the graph of the potential energy $P(\gamma, \sigma)$ is shown in Figure 8.6(b). The height

---

[5] In the theory of quasistatic processes of Chapter 3, it is shown that the dissipation inequality implies that nucleation from the low-strain phase to the high-strain phase cannot occur at a stress less than the Maxwell stress. We make no use of this result here.

$b = b(\sigma)$ between the low-strain phase minimum and the local maximum at $\gamma = \gamma_*$ is the *energy barrier* that a particle in the low-strain phase must surmount in order to jump to the high-strain phase at the stress σ. In particular, at the prenucleation stress σ_{pre}, the energy barrier can be computed with the help of (8.49), (8.54), and (8.55) to be

$$b(\sigma_{\text{pre}}) = P(\gamma_*, \sigma_{\text{pre}}) - P(\gamma_{\text{pre}}, \sigma_{\text{pre}}) = \frac{1}{2\mu}(\sigma_{\text{pre}} - \sigma_M)^2. \qquad (8.60)$$

Since σ_{pre} lies between σ_0 and σ_M, it follows from (8.60) that

$$0 < b(\sigma_{\text{pre}}) < \frac{1}{2\mu}(\sigma_M - \sigma_0)^2 = \mu\gamma_T^2/8. \qquad (8.61)$$

As σ_{pre} increases from σ_0 to σ_M, the barrier height $b(\sigma_{\text{pre}})$ decreases from $\mu\gamma_T^2/8$ to zero.

Assumption 2. *For nucleation to occur at a generic particle, it is necessary for the energy barrier $b(\sigma_{\text{pre}})$ to decrease to a value not greater than a materially determined critical value b_{nuc}, where $0 < b_{\text{nuc}} < \mu\gamma_T^2/8$.*

We thus take b_{nuc} to be given; it is convenient to write $b_{\text{nuc}} = \Sigma_{\text{nuc}}^2/(2\mu)$, where $\Sigma_{\text{nuc}} < \mu\gamma_T/2$, and to use Σ_{nuc} instead of b_{nuc} in the discussion to follow.

From Assumption 2 and (8.60), it follows that

$$\sigma_M \geq \sigma_{\text{pre}} \geq \sigma_M - \Sigma_{\text{nuc}} = \sigma_0 + \mu\gamma_T/2 - \Sigma_{\text{nuc}} > \sigma_0. \qquad (8.62)$$

In earlier chapters the nucleation criterion was phrased solely with reference to a single particle; this was possible because the nucleus of the new phase was taken to have vanishing size. In contrast here, the nucleus has a finite length s_{nuc} and so we must also consider the energetics of the *entire nucleus*. The nucleus $0 \leq x \leq s_{\text{nuc}}$ is itself effectively a bar; it is fixed at its left end $x = 0$ and is subjected to a loading on its right hand end $x = s_{\text{nuc}}$ by the rest of the bar. The total potential energy of the nucleus comprises the elastic energy of the nucleus and the energy of the loading device, the latter being equal to the elastic energy in the rest of the bar. Thus the potential energy of the nucleus equals the total energy in the bar.

We now wish to compare the total energy per unit cross-sectional area E_{post} in the bar after nucleation with the total energy E_{pre} before nucleation. In doing so, we shall generalize the model of the bar to include a constant surface energy[6] per unit area τ at the phase boundary when computing E_{post}. The surface energy accounts for local distortion very close to the interface between the phases[7]. Such distortion presumably occurs on a length scale that is $O(\tau/\mu) \ll L$. The total strain energy in the bar is found by integrating the strain energy per unit reference volume $W(\gamma) = P(\gamma) + \sigma\gamma$ over the length of the bar, making use of appropriate pre- and

---

[6] In order to determine the length s_{nuc} we must introduce a length scale parameter into the model by accounting for, for example, surface energy.
[7] For related discussion, see Section 52 of Christian [13].

post-nucleation versions of the strain and stress. One finds that

$$\begin{cases} E_{\text{pre}} = \sigma_{\text{pre}}^2 L/(2\mu), \\ E_{\text{post}} = E_{\text{pre}} + L\left[-(\sigma_{\text{pre}} - \sigma_0)\gamma_T \, s_{\text{nuc}}/L + (\mu/2)(\gamma_T s_{\text{nuc}}/L)^2\right] + \tau. \end{cases} \quad (8.63)$$

Assumption 3. *For nucleation, it is necessary that the total energy of the system should not increase:*

$$E_{\text{post}} \leq E_{\text{pre}}. \quad (8.64)$$

One easily shows that, for sufficiently small $(\tau/\mu)/L$, (8.64) holds provided

$$s_{\text{nuc}}^- \leq s_{\text{nuc}} \leq s_{\text{nuc}}^+, \quad (8.65)$$

where $s_{\text{nuc}}^{\mp} > 0$ are defined through

$$\gamma_T \, s_{\text{nuc}}^{\mp}/L = (\sigma_{\text{pre}} - \sigma_0)/\mu \mp \left\{[(\sigma_{\text{pre}} - \sigma_0)/\mu]^2 - 2(\tau/\mu)/L\right\}^{1/2}. \quad (8.66)$$

Assumption 4. *Nucleation occurs at the smallest value of σ_{pre} allowed by Assumption 2 and results in the nucleus of smallest length[8] s_{nuc} allowed by Assumption 3.*

Thus by (8.62) and (8.65),

$$\left.\begin{array}{r} \sigma_{\text{pre}} = \sigma_M - \Sigma_{\text{nuc}} = \sigma_0 + \mu\gamma_T/2 - \Sigma_{\text{nuc}}, \\ s_{\text{nuc}} = s_{\text{nuc}}^- = \left\{(\sigma_{\text{pre}} - \sigma_0)/\mu - \left\{[(\sigma_{\text{pre}} - \sigma_0)/\mu]^2 - 2(\tau/\mu)/L\right\}^{1/2}\right\}(L/\gamma_T). \end{array}\right\} \quad (8.67)$$

Moreover, the nucleation value of δ then follows from (8.54)$_1$, and (8.67)$_1$ as

$$\delta_{\text{nuc}}/L = \sigma_{\text{pre}}/\mu = \gamma_* - \Sigma_{\text{nuc}}/\mu. \quad (8.68)$$

For small values of $(\tau/\mu)/L$ and a fixed value of $\sigma_{\text{pre}} > \sigma_0$, one can obtain the approximate formula

$$s_{\text{nuc}} \sim \tau/(\gamma_T(\sigma_{\text{pre}} - \sigma_0)); \quad (8.69)$$

as expected, this makes it clear that $s_{\text{nuc}} = 0$ in the limit of zero surface energy. This, together with (8.54)$_2$, shows that, when surface energy is neglected, the present model reduces to the one used elsewhere in this book ($s_{\text{nuc}} = 0$, $\sigma_{\text{pre}} = \sigma_{\text{post}}$).

---

[8] As noted previously, the potential energy of a particle has to increase by the barrier height $b(\sigma_{\text{pre}})$ before it can move into the high-strain phase. Since in our model all of the particles in the segment $0 < x < s_{\text{nuc}}$ undergo the phase transition simultaneously, they must all increase their energies by $b(\sigma_{\text{pre}})$ during the process of transformation. Since this increase of energy is proportional to the length s_{nuc} of the nucleus, it is smaller the smaller s_{nuc} is. This is the motivation for requiring the nucleus to have its smallest allowable size.

Immediately after nucleation there is a phase boundary in the bar at $x = s_{\text{nuc}}$ that has the high-strain phase on its left. For the material with the stress–strain curve of Figure 8.6(a), the driving force on such a phase boundary is given by $f = \gamma_T(\sigma - \sigma_0)$. Just after nucleation, the driving force on the newly emerged phase boundary is thus $f_{\text{nuc}} = \gamma_T(\sigma_{\text{post}} - \sigma_0)$. From this and (8.54) and (8.67), one finds that

$$f_{\text{nuc}} = \gamma_T \left\{ (\mu\gamma_T/2 - \Sigma_{\text{nuc}})^2 - 2\mu\tau/L \right\}^{1/2}. \tag{8.70}$$

The stress σ_{pre} and the size s_{nuc} of the nucleus may be expressed in terms of f_{nuc}:

$$\left. \begin{aligned} \sigma_{\text{pre}} &= \sigma_0 + \left\{ (f_{\text{nuc}}/\gamma_T)^2 + 2\mu\tau/L \right\}^{1/2}, \\ s_{\text{nuc}} &= (L/\mu\gamma_T) \left[\left\{ (f_{\text{nuc}}/\gamma_T)^2 + 2\mu\tau/L \right\}^{1/2} - f_{\text{nuc}}/\gamma_T \right]. \end{aligned} \right\} \tag{8.71}$$

The nucleation condition can therefore be alternatively stated in terms of f_{nuc} and τ as follows: if the elongation δ is slowly increased from zero, nucleation occurs at a critical value $\delta = \delta_{\text{nuc}} = L\sigma_{\text{pre}}/\mu$ where σ_{pre} is given by $(8.71)_1$ and results in the emergence of a high-strain phase nucleus of size s_{nuc} given by $(8.71)_2$.

REFERENCES

[1] R. Abeyaratne, C. Chu, and R.D. James, Kinetics of materials with wiggly energies: theory and application to the evolution of twinning microstructures in a Cu–Al–Ni shape memory alloy. *Philosophical Magazine* A, **73** (1996), pp. 457–97.

[2] R. Abeyaratne and J.K. Knowles, Implications of viscosity and strain-gradient effects for the kinetics of propagating phase boundaries in solids. *SIAM Journal on Applied Mathematics*, **51** (1991), pp. 1205–21.

[3] R. Abeyaratne and J.K. Knowles, A continuum model of a thermoelastic solid capable of undergoing phase transitions. *Journal of the Mechanics and Physics of Solids*, **41** (1993), pp. 541–71.

[4] R. Abeyaratne and S. Vedantam, Propagation of a front by kink motion. in *IUTAM Symposium on Variations of Domains and Free-Boundary Problems in Solid Mechanics*, edited by P. Argoul et al., Solid Mechanics and its Applications, Volume 66. Kluwer, Dordrecht, 1999, pp. 77–84.

[5] R. Abeyaratne and S. Vedantam, A lattice-based model of the kinetics of twin boundary motion. *Journal of the Mechanics and Physics of Solids*, **51** (2003), pp. 1675–700.

[6] M. Achenbach, A model for an alloy with shape memory. *International Journal of Plasticity*, **5** (1989), pp. 371–95.

[7] M. Achenbach and I. Müller, Simulation of material behavior of alloys with shape memory. *Archive for Mechanics*, **37** (1985), pp. 573–85.

[8] W. Atkinson and N. Cabrera, Motion of a Frenkel–Kontorowa dislocation in a one–dimensional crystal. *Physical Review* A, **138** (1965), pp. 763–6.

[9] A.M. Balk, A.V. Cherkaev, and L.I. Slepyan, Dynamics of chains with non-monotone stress–strain relations. I. Model and numerical experiments. *Journal of the Mechanics and Physics of Solids*, **49** (2001), pp. 131–48.

[10] A.M. Balk, A.V. Cherkaev, and L.I. Slepyan, Dynamics of chains with non-monotone stress–strain relations. II. Nonlinear waves and waves of phase transition. *Journal of the Mechanics and Physics of Solids*, **49** (2001), pp. 149–71.

[11] K. Bhattacharya, Phase boundary propagation in heterogeneous bodies. *Proceedings of the Royal Society of London* A, **455** (1999), pp. 757–66.

[12] H.B. Callen, *Thermodynamics and an Introduction to Thermostatistics*, Chapter 14, Second edition. Wiley, New York, 1985.

[13] J.W. Christian, *The Theory of Transformations in Metals and Alloys, Part 1: Equilibrium and General Kinetic Theory*, Chapter 10, second edition. Pergamon Press, Oxford, 1975.

[14] J. Frenkel and T. Kontorowa, On the theory of plastic deformation and twinning. *Physikalische Zeitschrift der Sowjeturion*, **13** (1938), pp. 1–10.

[15] M. Grujicic, G.B. Olson, and W.S. Owen, Mobility of martensite interfaces. *Metallurgical Transactions*, **16A** (1985), pp. 1713–22.

[16] M. Grujicic, G.B. Olson, and W.S. Owen, Mobility of the $\beta_1 - \gamma_1$ martensitic interface in CuAlNi. Part I. Experimental measurements. *Metallurgical Transactions*, **16A** (1985), pp. 1723–34.

[17] K. Kadau, T.C. Germann, P.S. Lomdahl, and B.L. Holian, Microscopic view of structural phase transitions induced by shock waves. *Science*, **296** (2002), pp. 1681–4.

[18] J. Kestin, *A Course in Thermodynamics*, Volume II, Chapter 14, McGraw-Hill, New York, 1979.

[19] O. Kresse and L. Truskinovsky, Mobility of lattice defects: discrete and continuum approaches. *Journal of the Mechanics and Physics of Solids*, **51** (2003), pp. 1305–32.

[20] J.A. Krumhansl and J.R. Schrieffer, Dynamics and statistical mechanics of a one-dimensional model Hamiltonian for structural phase transitions. *Physical Review* B, **11** (1975), pp. 3535–45.

[21] P.G. LeFloch, *Hyperbolic Systems of Conservation Laws*. Birkhäuser Verlag, Basel, 2002.

[22] I. Müller and K. Wilmansky, Memory alloys - phenomenology and Ersatz-model, in *Continuum Models of Discrete Systems*, edited by O. Brulin and R.K.T. Hsieh. North-Holland, Amsterdam, 1981, pp. 495–509.

[23] D.A. Porter and K.E. Easterling, *Phase Transformations in Metals and Alloys*. van Nostrand-Reinhold, New York, 1981.

[24] J. Pouget, Dynamics of patterns in ferroelastic–martensitic transformations. I. Lattice model. *Physical Review* B, **43** (1991), pp. 3575–81.

[25] P.K. Purohit, *Dynamics of Phase Transitions in Strings, Beams and Atomic Chains*. Ph.D. dissertation, California Institute of Technology, Pasadena, CA, November 2001.

[26] G. Puglisi and L. Truskinovsky, Mechanics of a discrete chain with bistable elastic elements. *Journal of the Mechanics and Physics of Solids*, **48** (2000), pp. 1–27.

[27] M. Shearer, Nonuniqueness of admissible solutions of Riemann initial value problems for a system of conservation laws of mixed type. *Archive for Rational Mechanics and Analysis*, **93** (1986), pp. 45–59.

[28] M. Shearer, Dynamic phase transitions in a van der Waals gas. *Quarterly of Applied Mathematics*, **46** (1988), pp. 631–6.

[29] M. Slemrod, Admissibility criteria for propagating phase boundaries in a van der Waals fluid. *Archive for Rational Mechanics and Analysis*, **81** (1983), pp. 301–15.

[30] M. Slemrod, Dynamics of first order phase transitions, in *Phase Transitions and Material Instabilities in Solids*, edited by M.E. Gurtin. Academic Press, New York, 1984, pp. 163–203.

[31] M. Slemrod, A limiting "viscosity" approach to the Riemann problem for materials exhibiting change of phase. *Archive for Rational Mechanics and Analysis*, **105** (1989), pp. 327–65.

[32] S. Turteltaub, Viscosity and strain-gradient effects on the kinetics of propagating phase boundaries in solids. *Journal of Elasticity*, **46** (1997), pp. 53–90.

[33] L. Truskinovsky, Equilibrium phase interfaces. *Soviet Physics Doklady*, **27** (1982), pp. 551–3.

[34] L. Truskinovsky, Structure of an isothermal phase discontinuity. *Soviet Physics Doklady*, **30** (1985), pp. 945–8.

[35] L. Truskinovsky, Dynamics of non-equilibrium phase boundaries in a heat-conducting, non-linearly elastic medium. *Journal of Applied Mathematics and Mechanics (PMM, USSR)*, **51** (1987), pp. 777–84.

Part IV One-Dimensional Thermoelastic Theory and Problems

9 Models for Two-Phase Thermoelastic Materials in One Dimension

9.1 Preliminaries

Our objective in this chapter is to describe certain explicit and useful special constitutive models for two-phase thermoelastic materials in a one-dimensional setting. These models are formally applicable to either uniaxial stress in a bar or uniaxial strain in a slab. In the former case, one considers a thermoelastic bar occupying the interval $0 \leq x \leq L$ of the x-axis in a reference configuration that is stress free at the absolute temperature θ_0; in a motion, the particle at x in the reference configuration is carried at time t to $x + u(x, t)$. In the latter case, the slab occupies the region $0 \leq x \leq L$, $-\infty < y, z < \infty$ in the reference configuration, which is stress free at the temperature θ_0; the particle at x, y, z is carried to $x + u(x, t), y, z$ during a motion. In either case, we write

$$\gamma = u_x, \qquad v = u_t \tag{9.1}$$

for the strain and particle velocity, respectively. For a bar, the thermoelastic material is characterized by an internal energy per unit mass, or specific internal energy, $\hat{\epsilon}(\gamma, \eta)$, where η is the entropy per unit mass. For a thermoelastic slab, the internal energy per unit mass $\hat{\epsilon}(\mathbf{F}, \eta)$ for three-dimensional motions with deformation gradient tensor $\mathbf{F}$ and entropy per unit mass η reduces in uniaxial strain to a function of γ and η only. For the bar *or* the slab, the fundamental thermoelastic constitutive law yields

$$\sigma = \rho \hat{\epsilon}_\gamma(\gamma, \eta), \qquad \theta = \hat{\epsilon}_\eta(\gamma, \eta) > 0, \tag{9.2}$$

where ρ is the (constant) mass per unit reference volume, and subscripts γ and η indicate partial derivatives with respect to these quantities. In $(9.2)_1$, σ is the axial Piola–Kirchhoff normal stress, and θ is the absolute temperature.

In uniaxial strain, there are also transverse normal components of the Piola–Kirchhoff stress tensor, which in general do not coincide with σ and which cannot be calculated from the restriction of $\hat{\epsilon}(\mathbf{F}, \eta)$ to uniaxial strain. It is assumed that the symmetry of the material at the reference configuration is such as to admit uniaxial strain; this is true, for example, if the material is isotropic in the reference state.

Throughout this chapter, we shall for definiteness use the language of uniaxial stress in a bar, and in two-phase materials we shall assume the phase transition occurs in tension. Obvious modifications can be made, and would be needed, to deal with compression-induced phase changes, which are more natural from the experimental viewpoint. Useful background references for the material of this chapter are the article on thermoelasticity by Carlson [4] and the paper on shock waves in thermoelastic materials by Dunn and Fosdick [6].

For a thermoelastic material, the specific heat at constant strain $c(\gamma, \eta)$ and a measure of thermomechanical coupling called the *Grüneisen parameter*[1] are defined by

$$c = \frac{\hat{\epsilon}_\eta}{\hat{\epsilon}_{\eta\eta}}, \qquad \beta = -\frac{\hat{\epsilon}_{\gamma\eta}}{\hat{\epsilon}_\eta}. \tag{9.3}$$

We assume throughout that the specific heat c is positive, so that it follows from $(9.3)_1$ that $(9.2)_2$ is invertible to give

$$\eta = \hat{\eta}(\gamma, \theta). \tag{9.4}$$

The Helmholtz free energy per unit mass is then

$$\hat{\psi}(\gamma, \theta) = \hat{\epsilon}(\gamma, \hat{\eta}(\gamma, \theta)) - \theta\hat{\eta}(\gamma, \theta); \tag{9.5}$$

we may express the stress σ and the entropy η in the alternate forms

$$\sigma = \hat{\sigma}(\gamma, \theta) = \rho\hat{\psi}_\gamma(\gamma, \theta), \qquad \eta = \hat{\eta}(\gamma, \theta) = -\hat{\psi}_\theta(\gamma, \theta). \tag{9.6}$$

As functions of γ and θ, the specific heat at constant strain c and the Grüneisen parameter β are given by

$$c = -\theta\,\hat{\psi}_{\theta\theta}, \qquad \beta = \hat{\psi}_{\gamma\theta}/(\theta\hat{\psi}_{\theta\theta}). \tag{9.7}$$

The isentropic elastic modulus $\mu_*(\gamma, \eta) = \partial\sigma(\gamma, \eta)/\partial\gamma$ and the isothermal elastic modulus $\mu = \partial\hat{\sigma}(\gamma, \theta)/\partial\gamma$ are given by

$$\mu_* = \rho\hat{\epsilon}_{\gamma\gamma}, \qquad \mu = \rho\hat{\psi}_{\gamma\gamma}. \tag{9.8}$$

To relate the Grüneisen parameter β to a more familiar quantity, the coefficient of thermal expansion, we suppose for a moment that the bar is at a strain γ and a temperature θ, producing a stress $\sigma = \rho\hat{\psi}(\gamma, \theta)$. If the temperature is changed slightly to a value $\bar{\theta}$ while the stress is kept fixed at the value σ, one can approximately calculate the resulting strain $\bar{\gamma}$ by linearizing the stress–strain relation in the form $(9.6)_1$ about the state γ, θ. The result is $\bar{\gamma} = \gamma + \alpha(\bar{\theta} - \theta)$, where the *coefficient of thermal expansion α at the base state γ, θ* is defined by

$$\alpha = \rho c\beta/\mu; \tag{9.9}$$

---

[1] In the setting of uniaxial strain, the entity β defined by $(9.3)_2$ is often called the *modified* Grüneisen parameter, the unmodified version being $(1 + \gamma)\beta$.

here c, β, and μ are the respective values of the specific heat, the Grüneisen parameter, and the isothermal modulus at the base state.

Since the reference configuration is assumed to be unstressed at the temperature θ_0, one has

$$\hat{\epsilon}_\gamma(0, \eta_0) = 0, \quad \hat{\epsilon}_\eta(0, \eta_0) = \theta_0, \quad \text{or} \quad \hat{\psi}_\gamma(0, \theta_0) = 0, \quad \hat{\psi}_\theta(0, \theta_0) = -\eta_0, \quad (9.10)$$

where η_0 is the specific entropy in the reference configuration at the temperature θ_0. Without loss of generality, we may normalize the internal energy density by specifying its value at the reference configuration at the temperature θ_0; it is convenient to set

$$\hat{\epsilon}(0, \eta_0) = c(0, \eta_0)\theta_0, \qquad (9.11)$$

or equivalently $\hat{\psi}(0, \theta_0) = c(0, \eta_0)\theta_0 - \eta_0\theta_0$.

9.2 Materials of Mie–Grüneisen type

A class of thermoelastic models often referred to as *Mie–Grüneisen materials* has played a significant role in the study of shock waves in solids [11, 15]. Material models of this kind can be traced back to efforts early in the twentieth century to determine macroscopic properties of a solid such as its specific heat and coefficient of thermal expansion from statistical–mechanical considerations at the atomic level; see [8], Chapters 13 and 14 of [17], or Section 5.2 of [12].

From the continuum mechanical point of view, a thermoelastic material is of Mie–Grüneisen type if its specific heat at constant strain is independent of strain: $c(\gamma, \eta) = c(\eta)$. It is easy to show that this is the case if and only if the specific internal energy of the material is of the form

$$\hat{\epsilon}(\gamma, \eta) = p(\gamma)H(\eta) + q(\gamma), \qquad (9.12)$$

where $p(\gamma)$, $q(\gamma)$, and $H(\eta)$ are essentially arbitrary functions; $H(\eta)$ is related to $c(\eta)$ through

$$H'(\eta)/H''(\eta) = c(\eta). \qquad (9.13)$$

If $\hat{\epsilon}(\gamma, \eta)$ has the form (9.12), it follows that the Grüneisen parameter is independent of η:

$$\beta(\gamma, \eta) = \beta(\gamma); \qquad (9.14)$$

moreover, $p(\gamma)$ is related to $\beta(\gamma)$ through

$$p(\gamma) = \exp\left[-\int_0^\gamma \beta(\gamma')\right] d\gamma'. \qquad (9.15)$$

For a Mie–Grüneisen material, the constitutive law (9.2) specializes to

$$\sigma = -\rho\beta(\gamma)p(\gamma)H(\eta) + \rho q'(\gamma), \qquad \theta = p(\gamma)H'(\eta). \qquad (9.16)$$

Positivity of both temperature θ and specific heat c implies that $H'(\eta)$ is invertible; it then follows from $(9.16)_2$ that $\eta = K(\theta/p(\gamma))$, where K is the inverse of H'. From this, together with (9.5) and (9.12), one finds that the Helmholtz free energy density for a Mie–Grüneisen material must be of the form

$$\hat{\psi}(\gamma, \theta) = p(\gamma) L\left(\frac{\theta}{p(\gamma)}\right) + q(\gamma), \qquad (9.17)$$

for a suitable function L. We note in passing that the function of γ and θ obtained by adding an arbitrary linear function of temperature to the right side of (9.17) continues to have the form of a Helmholtz free energy density for a Mie–Grüneisen material.

That the Grüneisen parameter $\beta(\gamma, \eta)$ be independent of η is not only necessary for a thermoelastic material to be of Mie–Grüneisen type, it is also sufficient, as is easily shown. A thermoelastic material whose specific heat is constant, such as the ideal gas or the van der Waals fluid [10, 17], is a Mie–Grüneisen material.

In the special case of a Mie–Grüneisen material for which the specific heat at constant strain c and the modified Grüneisen parameter β are *both constants*, it is readily shown that (9.12) specializes to

$$\hat{\epsilon}(\gamma, \eta) = W(\gamma) + c\beta\,\theta_0\,\gamma + c\theta_0 \exp[(\eta - \eta_0)/c - \beta\gamma)], \qquad (9.18)$$

where we have set $W(\gamma) = q(\gamma) - c\beta\,\theta_0\,\gamma$. Using (9.10), (9.11), one finds that $W(\gamma)$ must satisfy

$$W(0) = W'(0) = 0, \qquad (9.19)$$

so that one has $W(\gamma) = O(\gamma^2)$ as $\gamma \to 0$. The constitutive expressions for stress and temperature that follow from (9.2) and (9.18) are

$$\sigma = \rho W'(\gamma) + \rho c\beta\,\theta_0\,\{1 - \exp[(\eta - \eta_0)/c - \beta\gamma]\},$$
$$\theta = \theta_0 \exp[(\eta - \eta_0)/c - \beta\gamma]. \qquad (9.20)$$

Using $(9.20)_2$, one may write $\eta = \hat{\eta}(\gamma, \theta) = \eta_0 + c\beta\,\gamma + c\log(\theta/\theta_0)$; by adding a suitable constant to η if necessary, we may assume that $\eta_0 = c$. The Helmholtz free energy density is then found according to (9.5) as

$$\hat{\psi}(\gamma, \theta) = W(\gamma) - c\beta(\theta - \theta_0)\gamma - c\theta \log(\theta/\theta_0). \qquad (9.21)$$

The associated representation of the constitutive law in terms of strain and temperature is

$$\sigma = \rho W'(\gamma) - \rho c\beta\,(\theta - \theta_0), \qquad \eta = c[1 + \beta\gamma + \log(\theta/\theta_0)]. \quad (9.22)$$

Upon linearization about the reference configuration, $(9.22)_1$ becomes

$$\sigma = \mu_0 \gamma - \rho c\beta(\theta - \theta_0), \qquad (9.23)$$

where $\mu_0 = \rho\hat{\psi}_{\gamma\gamma}(0, \theta_0) = \rho W''(0)$ is the referential isothermal elastic modulus; (9.23) is the classical stress–strain–temperature relation of one-dimensional linear thermoelasticity [4]. The coefficient of thermal expansion of the material is $\alpha = \rho c\beta/\mu_0$.

The *potential energy per unit reference volume P* is defined by

$$P(\gamma;\theta,\sigma) = \rho\hat{\psi}(\gamma,\theta) - \sigma\gamma; \qquad (9.24)$$

γ, θ, and σ are viewed as independent variables ranging over a suitable domain. Clearly

$$\frac{\partial P(\gamma;\theta,\sigma)}{\partial\gamma} = \hat{\sigma}(\gamma,\theta) - \sigma, \qquad \frac{\partial^2 P(\gamma;\theta,\sigma)}{\partial\gamma^2} = \frac{\partial\hat{\sigma}(\gamma,\theta)}{\partial\gamma}. \qquad (9.25)$$

It follows that, for fixed θ and σ, $P(\cdot;\theta,\sigma)$ has an extremum whenever (γ,σ) is a point on the isothermal stress–strain curve at the temperature θ, and that this extremum is a (local) minimum if the stress–strain curve has positive slope at the relevant point and a maximum if this slope is negative.

For Mie–Grüneisen materials whose Helmholtz free energy has the special form (9.21), one has

$$P(\gamma;\theta,\sigma) = \rho\, W(\gamma) - \rho c\beta(\theta - \theta_0)\gamma - \rho c\theta\log(\theta/\theta_0) - \sigma\,\gamma \qquad (9.26)$$

9.3 Two–phase Mie–Grüneisen materials

In order to characterize a particular thermoelastic material, it suffices to specify the Helmholtz free energy function $\hat{\psi}(\gamma,\theta)$ on a suitable region of the γ,θ-plane. For a *two-phase* thermoelastic material, many of the characteristics of $\hat{\psi}(\gamma,\theta)$ as a function of strain γ at a fixed temperature θ will be similar to those of the elastic strain energy encountered in the purely mechanical theory of Chapter 3.

In particular, for θ fixed in some range, the isothermal stress–strain curve associated with $\sigma = \hat{\sigma}(\gamma,\theta) = \rho\hat{\psi}_\gamma(\gamma,\theta)$ will rise as γ increases from -1 until it reaches a maximum at a point $(\gamma_M(\theta), \sigma_M(\theta))$, then decline to a minimum at $(\gamma_m(\theta), \sigma_m(\theta))$, and finally rise again. Precisely the same behavior occurs in the purely mechanical case, as illustrated in Figure 2.2. In the latter theory, the low- and high-strain phases correspond to the respective strain intervals $(-1, \gamma_M)$ and (γ_m, ∞). In the thermoelastic case, these phases are associated with regions P_L and P_H of the γ,θ-plane defined respectively by $-1 < \gamma < \gamma_M(\theta)$ and $\gamma > \gamma_m(\theta)$, with θ restricted to a suitable interval.

As in the purely mechanical case, for values of stress in the range $\sigma_m(\theta) < \sigma < \sigma_M(\theta)$, the potential energy function $P(\cdot;\theta,\sigma)$ for a two-phase material has two energy wells, one at $\gamma = \Gamma_L(\theta,\sigma)$ and the other at $\gamma = \Gamma_H(\theta,\sigma)$. The two energy wells have the same height when $\sigma = \sigma_0(\theta)$ – the Maxwell stress. The Gibbs free energies of the two phases are given by $g_L(\theta,\sigma) = P(\Gamma_L(\theta,\sigma);\theta,\sigma)$ and $g_H(\theta,\sigma) = P(\Gamma_H(\theta,\sigma);\theta,\sigma)$.

9.3.1 The trilinear material.
We now consider a thermoelastic material capable of existing in either of two phases, in each of which it coincides with a Mie–Grüneisen material with constant specific heat c and constant Grüneisen parameter β, the values of which we take to be the same in each phase. We characterize each phase

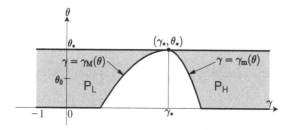

Figure 9.1. The regions, shown shaded, in the γ, θ-plane corresponding to the low-strain phase P_L and the high-strain phase P_H of the Mie–Grüneisen material characterized by the Helmholtz free energy density $\hat{\psi}(\gamma, \theta)$ of (9.27), and (9.28).

through its Helmholtz free energy density per unit mass, and we write

$$\hat{\psi}(\gamma, \theta) = \begin{cases} W(\gamma) - c\beta(\theta - \theta_0)\gamma - c\theta \log(\theta/\theta_0) \text{ for } (\gamma, \theta) \in P_L, \\[2mm] W(\gamma - \gamma_T) - c\beta(\theta - \theta_0)(\gamma - \gamma_T) - c\theta \log(\theta/\theta_0) + \lambda_T(\theta - \theta_0)/\theta_0 \\[2mm] \qquad\qquad\qquad \text{for } (\gamma, \theta) \in P_H, \end{cases}$$

(9.27)

where the regions P_L and P_H of the γ, θ-plane are associated with the low-strain and high-strain phases respectively. Here the constants $\gamma_T > 0$ and λ_T stand respectively for the *transformation strain* and the *latent heat of transformation* from the low-strain phase to the high-strain phase at the reference temperature θ_0. Moreover

$$\left. \begin{array}{l} P_L = \{(\gamma, \theta)| -1 < \gamma < \gamma_M(\theta), 0 < \theta < \theta_*\}, \\[2mm] P_H = \{(\gamma, \theta)| \gamma > \gamma_m(\theta), 0 < \theta < \theta_*\}, \end{array} \right\}$$

(9.28)

where the critical temperature θ_* satisfies $\theta_* > \theta_0$. The curves in the γ, θ-plane defined by $\gamma = \gamma_M(\theta)$ and $\gamma = \gamma_m(\theta)$, with $\gamma_M(\theta) < \gamma_m(\theta)$, $0 < \theta < \theta_*$, $\gamma_M(\theta_*) = \gamma_m(\theta_*)$, delineate the regions P_L and P_H that correspond to the two phases, as shown in Figure 9.1.

It is assumed that $W(\gamma)$ is strictly convex in γ and satisfies the normalization conditions (9.19); one then has $W(\gamma) \sim (\mu_0/(2\rho))\gamma^2$ as $\gamma \to 0$, so that $W(\gamma)$ has a minimum at $\gamma = 0$ and nowhere else. It follows from this and (9.27) that at the reference temperature θ_0, the Helmholtz free energy function $\hat{\psi}(\cdot, \theta_0)$ has two local minima: one at 0 and the other at γ_T. Moreover, observe that $\hat{\psi}(0, \theta_0) = \hat{\psi}(\gamma_T, \theta_0)$ so that the values of $\hat{\psi}$ at these two minima are the same at the temperature θ_0. Thus in the constitutive model (9.27), the reference temperature has been taken to be the transformation temperature: $\theta_0 = \theta_T$.

We shall use the term *trilinear thermoelastic material* [1] for the special case of (9.27) and (9.28) in which

$$W(\gamma) = \frac{\mu}{2\rho}\gamma^2, \qquad \left. \begin{array}{l} \gamma_M(\theta) = \gamma_* + M(\theta - \theta_*), \\[2mm] \gamma_m(\theta) = \gamma_* + m(\theta - \theta_*), \end{array} \right\}$$

(9.29)

where $\mu = \mu_0 > 0$, $\gamma_* > 0$, $\theta_* > 0$, and M and m are given material constants; in particular, the curves $\gamma = \gamma_M(\theta)$ and $\gamma = \gamma_m(\theta)$ shown in Figure 9.1 are now straight lines intersecting at the critical point γ_*, θ_*.

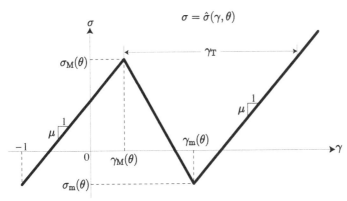

Figure 9.2. The isothermal trilinear stress–strain curve. The declining portion of the curve represents the unstable regime of the material.

The trilinear stress–strain–temperature relation that follows from (9.6) and (9.27) when (9.29) holds is

$$\sigma = \hat{\sigma}(\gamma, \theta) = \begin{cases} \mu\gamma - \rho c\beta(\theta - \theta_T) & \text{on } \mathsf{P_L}, \\ \mu(\gamma - \gamma_T) - \rho c\beta(\theta - \theta_T) & \text{on } \mathsf{P_H}. \end{cases} \tag{9.30}$$

At a fixed temperature θ, one has $\hat{\sigma}(-1, \theta) < \sigma < \sigma_M(\theta)$ in the low-strain phase $\mathsf{P_L}$, and $\sigma > \sigma_m(\theta)$ in the high-strain phase $\mathsf{P_H}$, where

$$\left.\begin{aligned} \sigma_M(\theta) &= \mu\gamma_M(\theta) - \rho c\beta(\theta - \theta_T), \\ \sigma_m(\theta) &= \mu(\gamma_m(\theta) - \gamma_T) - \rho c\beta(\theta - \theta_T), \end{aligned}\right\} \tag{9.31}$$

and $\gamma_M(\theta)$ and $\gamma_m(\theta)$ are given by $(9.29)_{2,3}$. We require that $-1 < \gamma_M(\theta) < \gamma_m(\theta)$ and $\sigma_M(\theta) > \sigma_m(\theta)$; these conditions hold for $0 < \theta < \theta_*$ if and only if the following inequalities are satisfied:

$$m < M < (1 + \gamma_*)/\theta_*, \qquad (M - m)\theta_* < \gamma_T. \tag{9.32}$$

A graph of the isothermal trilinear stress–strain curve is shown in Figure 9.2.

The line connecting the peak $(\gamma_M(\theta), \sigma_M(\theta))$ and the valley $(\gamma_m(\theta), \sigma_m(\theta))$ in the figure corresponds to the "unstable phase" of the material. If the constitutive law (9.6) is required to hold in the unstable regime $\gamma_M(\theta) \leq \gamma \leq \gamma_m(\theta)$, and if the stress–strain curve is piecewise linear and continuous as in the figure, then for $0 < \theta < \theta_*$, a simple geometric argument shows that

$$\rho\hat{\psi}(\gamma_m(\theta), \theta) - \rho\hat{\psi}(\gamma_M(\theta), \theta) = \tfrac{1}{2}[\sigma_M(\theta) + \sigma_m(\theta)][\gamma_m(\theta) - \gamma_M(\theta)]. \tag{9.33}$$

By using (9.27), (9.29), and (9.31), one can show that (9.33) holds for all $\theta \in (0, \theta_*)$ if and only if the following two relations are imposed on the various material constants:

$$\gamma_* = \gamma_T/2 + (M + m)(\theta_* - \theta_T)/2,$$

$$\lambda_T = (\mu \gamma_T \theta_T/\rho)[(M + m)/2 - \alpha].$$

(9.34)

The relation between specific entropy, strain, and temperature that follows from (9.6), (9.27), and (9.29) is

$$\eta = \hat{\eta}(\gamma, \theta) = \begin{cases} c\beta\, \gamma + c[1 + \log(\theta/\theta_T)] & \text{on } \mathsf{P_L}, \\ c\beta\, (\gamma - \gamma_T) + c[1 + \log(\theta/\theta_T)] - \lambda_T/\theta_T & \text{on } \mathsf{P_H}. \end{cases}$$

(9.35)

The latent heat of transformation $\lambda(\theta, \sigma)$ from $\mathsf{P_L}$ to $\mathsf{P_H}$ at the temperature θ and the stress σ is given by $\lambda = \theta\,[\hat{\eta}_L(\theta, \sigma) - \hat{\eta}_H(\theta, \sigma)]$ where $\hat{\eta}_L$ and $\hat{\eta}_H$ are the specific entropies associated with the two phases. Thus

$$\lambda = \lambda(\theta, \sigma) = \lambda_T \frac{\theta}{\theta_T}.$$

(9.36)

For simplicity and explicitness of results, the analysis in this and the next chapter will be carried out for the trilinear material subject to the restrictions (9.32) and (9.34); it will be clear, however, that the theory could be readily extended so as to apply to the more general two-phase Mie–Grüneisen material characterized by the Helmholtz free energy density (9.27), with $W(\gamma)$ assumed to be convex, though not quadratic, and to satisfy (9.19). The practical advantage of the trilinear material model arises from the fact that nonlinear effects are minimized.

9.3.2 Stability of phases of the trilinear material. Within each phase, the relation (9.30) can be inverted to give strain in terms of stress and temperature:

$$\left. \begin{aligned} \gamma &= \Gamma_L(\theta, \sigma) = \frac{\sigma}{\mu} + \alpha(\theta - \theta_T) \text{ on } \Pi_L, \\ \gamma &= \Gamma_H(\theta, \sigma) = \gamma_T + \frac{\sigma}{\mu} + \alpha(\theta - \theta_T) \text{ on } \Pi_H, \end{aligned} \right\}$$

(9.37)

where

$$\alpha = \rho c\beta/\mu$$

(9.38)

is the coefficient of thermal expansion of the material, and the regions Π_L and Π_H in the θ, σ-plane are the respective sets of states θ, σ at which the low-strain and high-strain phases can exist; they are defined by

$$\left. \begin{aligned} \Pi_L &= \{(\theta, \sigma)|\ \hat{\sigma}(-1, \theta) < \sigma < \sigma_M(\theta),\, 0 < \theta < \theta_*\}, \\ \Pi_H &= \{(\theta, \sigma)|\ \sigma > \sigma_m(\theta),\, 0 < \theta < \theta_*\}. \end{aligned} \right\}$$

(9.39)

The two phases can coexist on the region $\Pi_{LH} = \Pi_L \cap \Pi_H$. The regions Π_L, Π_H, and Π_{LH} in the θ, σ-plane are shown in Figure 9.3.

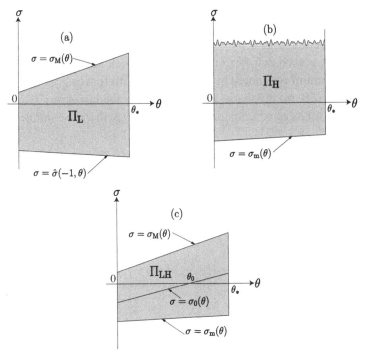

Figure 9.3. The low- and high-strain phases as represented by the respective regions Π_L (a) and Π_H (b) in the θ, σ-plane; Π_{LH} (c) is the region where the two phases may coexist. The *Maxwell line* $\sigma = \sigma_0(\theta)$ shown in (c) goes through the states (γ, θ) for which the Gibbs free energies of the two phases coincide; see (9.43). The slope of the Maxwell line is positive, as is the case in the figure, when $\lambda_T > 0$; see (9.44).

We define the *Gibbs free energy* per unit reference volume for each phase by

$$\left. \begin{aligned} g_L(\theta, \sigma) &= \rho \, \hat{\psi}(\Gamma_L(\theta, \sigma), \theta) - \sigma \, \Gamma_L(\theta, \sigma), \\ g_H(\theta, \sigma) &= \rho \, \hat{\psi}(\Gamma_H(\theta, \sigma), \theta) - \sigma \, \Gamma_H(\theta, \sigma), \end{aligned} \right\} \quad \text{on } \Pi_{LH}. \qquad (9.40)$$

Now consider the potential energy $P(\gamma; \theta, \sigma)$ of (9.24) specialized through (9.27), and (9.29) to the two-phase trilinear material. Let (θ, σ) be fixed in Π_L, and seek the extrema of $P(\gamma; \theta, \sigma)$ as γ varies over the interval $(-1, \gamma_M(\theta))$. At such an extremum, (9.25)$_1$ shows that $\gamma = \Gamma_L(\theta, \sigma)$ must hold; since W is convex, there is exactly one extremum, and it is a minimum. By (9.24) and (9.40), the value of P at the minimum is precisely the Gibbs free energy density for the low-strain phase: for $(\theta, \sigma) \in \Pi_L$,

$$\min_{\gamma \in (-1, \gamma_M(\theta))} P(\gamma; \theta, \sigma) = P(\Gamma_L(\theta, \sigma); \theta, \sigma) = g_L(\theta, \sigma). \qquad (9.41)$$

Similarly, if (θ, σ) is fixed in the high-strain phase Π_H, the minimum value of the potential energy as γ varies over $(\gamma_m(\theta), \infty)$ is the Gibbs free energy for the high-strain phase $g_H(\theta, \sigma)$.

Parenthetically, it may be noted that for (θ, σ) fixed in the coexistence region Π_{LH}, as γ ranges over the interval $(\gamma_M(\theta), \gamma_m(\theta))$ corresponding to the unstable regime described by the declining branch of the curve in Figure 9.2, $P(\gamma; \theta, \sigma)$ has a *maximum* at the value of γ such that (γ, σ) lies on this branch for the given θ. If the value of P at this maximum is denoted by $g_U(\theta, \sigma)$, the differences $g_U(\theta, \sigma) - g_L(\theta, \sigma)$ and $g_U(\theta, \sigma) - g_H(\theta, \sigma)$ are the *energy barriers* for the low-strain – to – high-strain transition and the reverse transition, respectively, at the temperature θ and stress σ.

In order to determine which of the two phases is energetically preferred, or "stable", when both are available, we now compare the two minima $g_L(\theta, \sigma)$ and $g_H(\theta, \sigma)$ when (θ, σ) lies in the coexistence region Π_{LH}. Using (9.27), (9.29), and (9.37), we obtain

$$\left.\begin{array}{l} g_L(\theta, \sigma) = -\frac{1}{2\mu}[\sigma + \mu\alpha(\theta - \theta_T)]^2 - \rho c\theta \log(\theta/\theta_T) \quad \text{on } \Pi_L, \\[2mm] g_H(\theta, \sigma) = -\frac{1}{2\mu}[\sigma + \mu\alpha(\theta - \theta_T)]^2 - \sigma\gamma_T - \rho c\theta \log(\theta/\theta_T) + \rho\lambda_T\frac{\theta - \theta_T}{\theta_T} \quad \text{on } \Pi_H. \end{array}\right\}$$

(9.42)

From (9.42), we find the difference $g_L - g_H$ as

$$g_L(\theta, \sigma) - g_H(\theta, \sigma) = \gamma_T[\sigma - \sigma_0(\theta)] \quad \text{for } (\theta, \sigma) \in \Pi_{LH}, \qquad (9.43)$$

where

$$\sigma_0(\theta) = \frac{\rho\lambda_T}{\gamma_T}\left(\frac{\theta - \theta_T}{\theta_T}\right). \qquad (9.44)$$

The stress $\sigma_0(\theta)$ at which the Gibbs free energies of the two phases coincide is the *Maxwell stress* at the temperature θ. The formula (9.43) may be compared with its counterpart for the trilinear, two-phase elastic material in the purely mechanical theory of mixed phase equilibria; see equations (3.5) and (3.53). By appealing to (9.31), (9.34), and (9.29)$_{2,3}$, one can show that

$$\sigma_0(\theta) = \tfrac{1}{2}[\sigma_M(\theta) + \sigma_m(\theta)], \qquad (9.45)$$

just as in the purely mechanical theory of trilinear elastic bars in Chapter 3.

Because of (9.45), it is clear that the Maxwell line $\sigma = \sigma_0(\theta)$ in the θ, σ-plane lies in the coexistence region Π_{LH} of Figure 9.3. It then follows from (9.43) that, at states (θ, σ) that lie strictly above the Maxwell line, the high-strain phase minimum of the potential energy P lies below its low-strain phase minimum, so that the high-strain phase is preferred, or stable; while below the Maxwell line, the low-strain phase is stable. *On the Maxwell line, the values of P at its two minima coincide, and both phases are stable and in phase equilibrium.* In particular, this is true of the unstressed state ($\sigma = 0$) at the temperature θ_T reflecting the fact that θ_T is the *transformation temperature*.

The referential latent heat λ_T will be positive if $(M + m)\theta_T/2 > \alpha\theta_T$.

The trilinear two-phase thermoelastic material described by (9.27), (9.29), (9.30) and (9.35) is unlikely to yield reliable results near the critical point (γ_*, θ_*) in the γ, θ-plane; see Section 20.4 of [10] or Chapter 10 of [3]. Neither is there reason to expect the trilinear model to be realistic near $\theta = 0$. Indeed, the fact that the thermomechanical coupling in (9.30) is linear in $\theta - \theta_T$ suggests that the applicability of the model is likely to be limited to temperatures near the transformation temperature θ_T.

The trilinear model was developed to describe thermoelastic martensitic transformations. Suppose that the tensile bar composed of a trilinear material is initially in the stress-free reference configuration at the transformation temperature θ_T, and assume that the referential latent heat λ_T is positive. If the bar is now heated to a temperature $\theta > \theta_T$, reference to Figure 9.3 shows that the bar will remain in the low-strain phase Π_L, which is stable. If instead the bar were cooled to a temperature less than θ_T, it would find itself in the high-strain phase Π_H, which is the stable one for $\theta < \theta_T$, $\sigma = 0$. In keeping with the terminology of materials science, we would then call the low-strain phase *austenite*, the high-strain phase *martensite*. The names would be reversed if the latent heat λ_T were negative. Although in real materials martensite often exists in many variants, the present trilinear model acknowledges only one.

9.3.3 Other two-phase materials of Mie–Grüneisen type.

We shall briefly describe other thermoelastic models of Mie–Grüneisen type that have been used in the study of phase transitions in solids.

Certain experiments in which a glass specimen is compressed by impact have been interpreted as involving a "failure wave" that follows a front-running shock wave; see, for example, [2]. Behind the slowly moving failure wave, the glass is thought to be in a nearly comminuted state, with greatly reduced strength. Clifton [5] investigated a model in which the failure wave is viewed as a propagating phase boundary associated with a phase transition involving large strains that promote stress-induced microcracking behind the failure wave. The analysis in [5] treats the glass as a Mie–Grüneisen material whose specific heat at constant strain c and Grüneisen parameter β are constants and whose internal energy density is therefore of the form (9.18). The function $W(\gamma)$ chosen by Clifton has the form

$$W(\gamma) = \frac{\mu}{2\rho}(\gamma^2 + a\gamma^3 + b\gamma^4), \qquad (9.46)$$

where a and b are positive constants. If a and b are suitably restricted, this $W(\gamma)$ corresponds to a Mie–Grüneisen model for a material that changes phase in compression and leads through (9.22) and (9.20) to nonmonotonic isothermal and isentropic stress–strain curves of the form illustrated schematically in Figure 9.4.

Motivated by the possibility of large strains arising from phase transitions in the earth's mantle, Sammis and Dein [16] carried out a laboratory experiment designed to show that creep can be enhanced by the occurrence of a phase transition induced by combined thermal and mechanical loads. A model for their experiment given

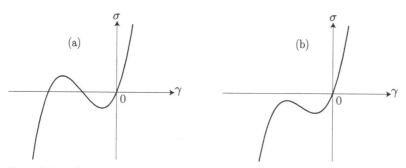

Figure 9.4. Isothermal and isentropic stress–strain curves for a two-phase Mie–Grüneisen model based on $(9.22)_1$ and $(9.20)_1$, with $W(\gamma)$ of the form (9.46). (a) An isothermal stress–strain curve at the referential temperature $\theta = \theta_T$. (b) An isentropic stress–strain curve at the referential specific entropy $\eta = \eta_0$.

by Jiang and Knowles [9] involves a two-phase thermoelastic material of Mie–Grüneisen type whose Helmholtz free energy density is of the form

$$\hat{\psi}(\gamma, \theta) = \frac{\mu}{2\rho}\left(\gamma^2 - a\frac{\theta - \theta_*}{\theta_T - \theta_*}\gamma^4 + \frac{1}{3}\gamma^6\right) - c\theta\log(\theta/\theta_T). \quad (9.47)$$

In the relevant physical context, γ is a shear strain, a is a positive constant, θ_* and $\theta_T > \theta_*$ are the critical temperature and the transformation temperature, respectively, and c is the constant specific heat at constant strain. The Helmholtz free energy (9.47) is a special case of one introduced by Falk [7] in his study of shape memory alloys and also used by Müller and Xu [13] to investigate pseudoelastic hysteresis. In essence, Falk's free energy differs from that in (9.47) only in that the term $c\theta\log(\theta/\theta_T)$ in (9.47) is replaced by an arbitrary function of θ in [7].

For any fixed temperature $\theta < \theta_*$, $\hat{\psi}(\cdot, \theta)$ of (9.47) is convex with a minimum at $\gamma = 0$, and the material can exist in one phase only, regardless of the value of the shear stress $\sigma = \rho\hat{\psi}_\gamma$; see Figure 9.5. For temperatures above the critical temperature, $\hat{\psi}(\cdot, \theta)$ is nonconvex. There is a temperature $\bar{\theta}$ intermediate between θ_* and θ_T such that for $\theta_* < \theta < \bar{\theta}$, $\hat{\psi}(\cdot, \theta)$ still has only the single well at $\gamma = 0$, so there is only one unstressed state. On the other hand, for $\theta > \bar{\theta}$, $\hat{\psi}(\cdot, \theta)$ has a minimum at $\gamma = 0$ and two additional minima symmetrically placed with respect to $\gamma = 0$, so that there are three unstressed states. If $\bar{\theta} < \theta < \theta_T$, the value of $\hat{\psi}$ at the minimum at $\gamma = 0$ lies below that for the other minima, so the unsheared state is the stable stress-free state. For $\theta > \theta_T$, the two sheared stress-free states, at which the values of $\hat{\psi}$ are the same, have the lower minima, and are therefore the stable states. At the transformation temperature $\theta = \theta_T$, all three stress-free states are stable. Figure 9.5 shows graphs of $\hat{\psi}(\gamma, \theta)$ as a function of γ for various values of θ.

The Grüneisen parameter β arising from the Helmholtz free energy density $\hat{\psi}$ of (9.47) is found from (9.7) to be

$$\beta = \beta(\gamma) = -(1/c)\hat{\psi}_{\gamma\theta}(\gamma, \theta) = \frac{4a}{c(\theta_T - \theta_*)}\gamma^3; \quad (9.48)$$

note that $\beta(\gamma)$ vanishes in the reference configuration.

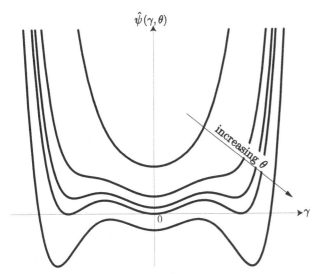

Figure 9.5. Helmholtz free energy $\hat{\psi}(\gamma, \theta)$ of (9.47) as a function of strain γ at various temperatures θ, with temperature increasing in the indicated direction. For temperatures lower than the transformation temperature θ_T, the unsheared state has minimum energy, while for $\theta > \theta_T$, the sheared states are preferred.

In their studies of propagating phase boundaries, Turteltaub [18] and Ngan and Truskinovsky [14] make use of Helmholtz free energies of the Mie–Grüneisen form (9.21). In [18], the two-phase thermoelastic material is trilinear, whereas in [14], $W(\gamma)$ is a quartic polynomial that characterizes a two-phase material. In both of these works, the associated thermoelastic constitutive law is augmented by means of viscosity and strain gradient effects, and the kinetic relation inherited by the limiting sharp-interface theory through this augmentation is determined.

REFERENCES

[1] R. Abeyaratne and J.K. Knowles, A continuum model of a thermoelastic material capable of undergoing phase transitions. *Journal of the Mechanics and Physics of Solids*, **41** (1993), pp. 541–71.

[2] N.S. Brar, S.J. Bless, and Z. Rosenberg, Impact-induced failure waves in glass bars and plates. *Applied Physics Letters*, **59** (1991), pp. 3396–8.

[3] H.B. Callen, *Thermodynamics and an Introduction to Thermostatistics*, second edition. Wiley, New York, 1985.

[4] D.E. Carlson, Linear thermoelasticity, in *Handbuch der Physik*, edited by C. Truesdell, Volume VIa/2. Springer, Berlin, 1972, pp. 297–345.

[5] R.J. Clifton, Analysis of failure waves in glasses. *Applied Mechanics Reviews*, **46** (1993), pp. 540–6.

[6] J.E. Dunn and R.L. Fosdick, Steady, structured shock waves. Part 1: Thermoelastic materials. *Archive for Rational Mechanics and Analysis*, **104** (1988), pp. 295–365.

[7] F. Falk, Model free energy, mechanics and thermodynamics of shape-memory alloys. *Acta Metallurgica*, **28** (1980), pp. 1773–80.

[8] E. Grüneisen, Zustand des festen Körpers, in *Handbuch der Physik*, edited by H. Geiger and K. Scheel, volume 10, Springer, Berlin, 1926.

[9] Q. Jiang and J.K. Knowles, A thermoelastic model for an experiment involving a solid–solid phase transition. *Continuum Mechanics and Thermodynamics*, **7** (1995), pp. 97–110.

[10] J. Kestin, *A Course in Thermodynamics*, Volume 2. Hemisphere Publishing, McGraw-Hill, Washington, DC, 1979.

[11] R.G. McQueen, S.P. Marsh, J.W. Taylor, J.N. Fritz, and W.J. Carter, The equation of state of solids from shock wave studies, in *High-Velocity Impact Phenomena*, edited by R. Kinslow. Academic Press, New York, 1970, pp. 294–417.

[12] M.A. Meyers, *Dynamic Behavior of Materials*. Wiley Interscience, New York, 1994.

[13] I. Müller and H. Xu, On the pseudo-elastic hysteresis. *Acta Metallurgica et Materialia*, **39** (1991), pp. 263–271.

[14] S.C. Ngan and L. Truskinovsky, Thermal trapping and the kinetics of martensitic phase boundaries. *Journal of the Mechanics and Physics of Solids*, **47** (1999), pp. 141–72.

[15] M.H. Rice, R.G. McQueen, and J.L. Walsh, Compression of solids by strong shock waves, in *Solid State Physics*, edited by F. Seitz and D. Turnbull, Volume 6. Academic Press, New York, 1958, pp. 1–63.

[16] C.G. Sammis and J.L. Dein, On the possibility of transformational superplasticity in the earth's mantle. *Journal of Geophysical Research*, **79** (1974), pp. 2961–5.

[17] J.C. Slater, *Introduction to Chemical Physics*. McGraw-Hill, New York, 1939.

[18] S. Turteltaub, Viscosity and strain-gradient effects on the kinetics of propagating phase boundaries in solids. *Journal of Elasticity*, **46** (1997), pp. 53–90.

10 Quasistatic Hysteresis in Two-Phase Thermoelastic Tensile Bars

The purely mechanical quasistatic response of one-dimensional, two-phase elastic bars was discussed in Chapter 3. In the present chapter, we shall generalize that discussion to incorporate thermal effects. After setting out some preliminaries in Section 10.1, in Section 10.2 we describe the thermomechanical equilibrium states of a two-phase material. Quasistatic processes, taken to be one-parameter families of equilibrium states, are studied in Section 10.3. We specialize the discussion to a trilinear thermoelastic material in Section 10.4 and then evaluate the response of the bar to some specific loading programs: in Sections 10.5, 10.6, and 10.7 we consider stress cycles at constant temperature, temperature cycles at constant stress, and the shape-memory cycle respectively; qualitative comparisons with some experiments are also made in these sections. In Section 10.8 we describe an experimental result of Shaw and Kyriakides [15, 16] and compare the theoretical predictions of our model with it. Finally in Section 10.9 we comment on processes that are slow in the sense that inertial effects can be neglected but are not quasistatic in the preceding sense of being one-parameter families of equilibrium states.

10.1 Preliminaries

We begin by setting out the one-dimensional version of the theory of thermoelasticity given in Chapter 7. Consider a tensile bar that occupies the interval $[0, L]$ of the x-axis in a reference configuration. Let the bar undergo a longitudinal motion, so that the particle at x in the reference configuration is carried to the point $x + u(x, t)$ at time t and bears the temperature $\theta(x, t)$. It is assumed that u is continuous in (x, t) with piecewise continuous first and second partial derivatives for all x in $[0, L]$ and for all times t of interest, and that $\theta(\cdot, t)$ is piecewise continuous with piecewise continuous partial derivative on $[0, L]$. The strain and particle velocity at the particle x at time t are given by $\gamma(x, t) = u_x(x, t)$ and $v(x, t) = u_t(x, t)$ where the subscripts denote partial differentiation.

For a thermoelastic material in one dimension, the nominal stress $\sigma(x, t)$ and the specific entropy $\eta(x, t)$ are given by the thermoelastic constitutive law

$$\sigma = \rho\, \psi_\gamma(\gamma, \theta), \qquad \eta = -\psi_\theta(\gamma, \theta), \tag{10.1}$$

where $\psi(\gamma, \theta)$ is the Helmholtz free energy per unit mass, and ρ is the referential mass density. The heat flux $q(x, t)$ is taken to be given by the heat conduction law

$$q = K(\gamma, \theta)\,\theta_x, \tag{10.2}$$

where $K(\gamma, \theta)$ is the thermal conductivity. It is assumed that the material of the bar is homogeneous in the reference state, so that ψ, ρ, and K do not depend explicitly on x.

Wherever the fields are smooth, kinematic compatibility and the one-dimensional versions of the local balance laws (6.4) and (6.6) require that

$$v_x - \gamma_t = 0, \tag{10.3}$$

$$\sigma_x = \rho v_t, \tag{10.4}$$

$$q_x = \rho\theta\eta_t, \tag{10.5}$$

where we have neglected the effects of body force and heat supply. If there is a discontinuity in the fields at $x = s(t)$ consistent with the assumed smoothness, the jump conditions (7.53)–(7.55) require that

$$[\![\gamma]\!]\,\dot{s} + [\![v]\!] = 0, \tag{10.6}$$

$$[\![\sigma]\!] + \rho[\![v]\!]\,\dot{s} = 0, \tag{10.7}$$

$$(\rho[\![\epsilon]\!] - \langle\sigma\rangle\,[\![\gamma]\!])\dot{s} + [\![q]\!] = 0. \tag{10.8}$$

Finally, the entropy inequality is satisfied by taking $K > 0$ and requiring that

$$f\dot{s} \geq 0 \quad \text{where} \quad f = [\![\rho\psi]\!] - \langle\sigma\rangle\,[\![\gamma]\!] + \langle\rho\eta\rangle\,[\![\theta]\!] \tag{10.9}$$

is the driving force; see (6.28) and (6.33).

Now consider the special case of *thermomechanical equilibrium* and suppose that (i) the fields u and θ are independent of time and (ii) the temperature is continuous. Then $v = 0$, and by the thermoelastic constitutive law (10.1) and the heat conduction law (10.2), all fields are independent of time. The field equations and interface conditions (10.3)–(10.9) now require that the stress σ and the heat flux q be uniform through the bar, and the driving force may be expressed as

$$f = [\![\rho\psi]\!] - \sigma[\![\gamma]\!]. \tag{10.10}$$

10.2 Thermomechanical equilibrium states for a two-phase material

Suppose that the end $x = 0$ of the bar is fixed, while the end $x = L$ is given a prescribed axial displacement $\bar{\delta}$. Furthermore assume that the value of temperature is prescribed to be $\theta = \bar{\theta}$ at each end of the bar. Thus as boundary conditions, we have

$$u(0) = 0, \qquad u(L) = \bar{\delta}, \qquad \text{and} \quad \theta(0) = \theta(L) = \bar{\theta}. \qquad (10.11)$$

We wish to find $u(x)$, $\gamma(x)$, $\theta(x)$, $\eta(x)$, σ, and q such that (10.1), (10.2), and (10.11) are satisfied; we speak of this as the *equilibrium problem*.

It is clear that

$$u(x) = (\bar{\delta}/L)x, \qquad \gamma(x) = \bar{\delta}/L, \qquad \theta(x) = \bar{\theta} \quad \text{for } 0 \leq x \leq L \qquad (10.12)$$

along with

$$\sigma = \rho\psi_\gamma(\bar{\delta}/L, \bar{\theta}), \qquad \eta = -\psi_\theta(\bar{\delta}/L, \bar{\theta}), \qquad q = 0 \quad \text{for } 0 \leq x \leq L \qquad (10.13)$$

comprises a solution of the equilibrium problem. In the absence of restrictions on ψ, there is no reason to assume that this is the only solution. In particular, for a two-phase thermoelastic material, we should expect from our discussion of the corresponding problem for the elastic bar in Chapter 3 that there will, in general, be other solutions.

Suppose now that the material is a general two-phase thermoelastic material as introduced in the preceding chapter. For such a material, the Helmholtz free energy function of the two phases can be characterized by

$$\psi(\gamma, \theta) = \begin{cases} \psi_L(\gamma, \theta) \text{ for } (\gamma, \theta) \in P_L, \\[2mm] \psi_H(\gamma, \theta) \text{ for } (\gamma, \theta) \in P_H, \end{cases} \qquad (10.14)$$

where the regions P_L and P_H in the γ, θ-plane correspond to the low- and high-strain phases respectively; in the special case of a trilinear material these regions are defined by (9.28). The corresponding stress and entropy are given by

$$\sigma = \rho\psi_\gamma(\gamma, \theta) = \begin{cases} \sigma_L(\gamma, \theta) \text{ for } (\gamma, \theta) \in P_L, \\[2mm] \sigma_H(\gamma, \theta) \text{ for } (\gamma, \theta) \in P_H, \end{cases} \qquad (10.15)$$

$$\eta = -\psi_\theta(\gamma, \theta) = \begin{cases} \eta_L(\gamma, \theta) \text{ for } (\gamma, \theta) \in P_L, \\[2mm] \eta_H(\gamma, \theta) \text{ for } (\gamma, \theta) \in P_H. \end{cases} \qquad (10.16)$$

For each individual phase, the stress–strain–temperature relation can be inverted to express the strain as a function of stress and temperature:

$$\gamma = \begin{cases} \Gamma_L(\theta, \sigma) \text{ for } (\theta, \sigma) \in \Pi_L, \\[2mm] \Gamma_H(\theta, \sigma) \text{ for } (\theta, \sigma) \in \Pi_H, \end{cases} \qquad (10.17)$$

where the regions Π_L and Π_H in the θ, σ-plane correspond to the low- and high-strain phases respectively; in the special case of a trilinear material these regions are defined by (9.39). When (θ, σ) belongs to $\Pi_{HL} = \Pi_L \cap \Pi_H$, the material can exist in either phase.

For notational simplicity, we now drop the overbars from the prescribed elonga-
tion $\bar{\delta}$ and temperature $\bar{\theta}$. Suppose that the prescribed temperature θ, and the stress
σ associated with a solution to the equilibrium problem, turn out to correspond to
a point in the coexistence region Π_{LH}. We now seek a mixed-phase solution of the
problem of the form

$$\gamma(x) = \begin{cases} \gamma^- & \text{for} \quad 0 < x < s, \\ \\ \gamma^+ & \text{for} \quad s < x < L, \end{cases} \qquad \theta(x) = \theta \quad \text{for} \quad 0 < x < L, \quad (10.18)$$

in which (γ^-, θ) lies in the high-strain phase region P_H, (γ^+, θ) lies in the low-
strain phase region P_L, and the interface at $x = s$ is a phase boundary. The strains
$\gamma^\pm$ are given by (10.17) to be

$$\gamma^- = \Gamma_H(\theta, \sigma), \qquad \gamma^+ = \Gamma_L(\theta, \sigma), \qquad (10.19)$$

and so the overall elongation $\delta = \gamma^- s + \gamma^+(L - s)$ of the bar is

$$\delta = \Gamma_H(\theta, \sigma)s + \Gamma_L(\theta, \sigma)(L - s). \qquad (10.20)$$

By using (10.19) we find from (10.10) that the driving force can be expressed
as

$$f = \rho\psi(\Gamma_L(\theta, \sigma), \theta) - \rho\psi(\Gamma_H(\theta, \sigma), \theta) - \sigma\left[\Gamma_L(\theta, \sigma) - \Gamma_H(\theta, \sigma)\right], \quad (10.21)$$

or, in terms of the Gibbs free energies of the two material phases as defined in
Section 9.3.2,

$$f = g_L(\theta, \sigma) - g_H(\theta, \sigma). \qquad (10.22)$$

The driving force coincides with the jump in the Gibbs free energy across the phase
boundary in the present setting.

10.3 Quasistatic processes

We now consider quasistatic processes of the bar, by which we mean one-parameter
families of mixed-phase equilibria of the kind just described, the parameter being
time t. This is entirely parallel to the discussion of quasistatic processes in elastic
bars in Chapter 3. Thus we now take $\sigma = \sigma(t)$, $\theta = \theta(t)$, $\delta = \delta(t)$, $s = s(t)$, and so
on in the equilibrium theory above.

As before, the entropy inequality $f(t)\dot{s}(t) \geq 0$ shows that the direction of propagation of the phase boundary, and therefore which phase grows at the expense of
the other, is determined by the sign of the driving force. In view of (10.22), the
phase favored for growth in quasistatic processes will be the one with the smaller
Gibbs free energy. If these two energies are equal, then the driving force vanishes,
and $(10.9)_1$ permits the phase boundary to move in either direction. In this case, the
quasistatic process is said to take place *reversibly*.

The evolution of the phase transformation is governed by a kinetic law

$$\dot{s} = \Phi(f, \theta), \qquad (10.23)$$

where the kinetic response function $\Phi(f, \theta)$ depends on the material. The entropy inequality requires that

$$\Phi(f, \theta)f \geq 0. \tag{10.24}$$

Having in mind the processes commonly studied in experiments, we suppose specifically that the histories of the temperature $\theta(t)$ and one of the two mechanical variables $\delta(t)$ or $\sigma(t)$ are prescribed; the history of the remaining mechanical variable is to be determined.

The basic system controlling the evolution of the thermomechanical state of the bar when both phases are present is then determined from (10.20), (10.23), and (10.22) as

$$\left. \begin{array}{l} \dot{s} = \Phi(g_L(\theta, \sigma) - g_H(\theta, \sigma), \theta), \\[2mm] \delta = \Gamma_L(\theta, \sigma)s + \Gamma_H(\theta, \sigma)(L - s). \end{array} \right\} \tag{10.25}$$

Among the four quantities s, θ, σ, and δ entering (10.25), two will be prescribed as functions of time (θ and δ, or θ and σ), while the remaining two are to be determined[1]. In an isothermal constant-elongation-rate process of the kind often carried out in the laboratory, one specifies constant values for $\theta(t)$ and $\dot{\delta}(t)$, and $\sigma(t)$ is to be found. On the other hand, in thermal cycling experiments, one specifies constant values for $\dot{\theta}(t)$ and $\sigma(t)$ (usually $\sigma = 0$), and one wishes to find $\delta(t)$.

Finally we turn to nucleation. In the quasistatic processes to be encountered here, the bar will be in the low-strain phase at the beginning of the thermomechanical loading process, and will remain there until nucleation of the transition to the high-strain phase takes place. As in the corresponding theory of quasistatic processes for elastic bars in Chapter 3, we assume that the end $x = 0$ of the bar is the nucleation site, and that the nucleation criterion specifies a critical value of driving force $f_n^{LH}(\theta) \geq 0$:

$$f \geq f_n^{LH}(\theta) \quad \text{(nucleation, low-strain to high-strain transition).} \tag{10.26}$$

After nucleation, the phase boundary moves to the right under the control of the kinetic relation; if and when it arrives at the end $x = L$, the bar is fully transformed to the high-strain phase. If the loading is subsequently reversed, nucleation of the reverse transition occurs if $f \leq f_n^{HL}(\theta) \leq 0$; the two nucleation values of driving force depend, in general, on the material and the temperature.

10.4 Trilinear thermoelastic material

In order to calculate the thermomechanical response of the bar explicitly, we now specialize the general formulation above to a trilinear Mie–Grüneisen material. For

---

[1] One can eliminate s from (10.25) to obtain a single master differential equation involving δ, σ, and θ only.

such a material, we have from (9.37) and (9.42) that

$$\Gamma_H(\theta, \sigma) = \gamma_T + \frac{\sigma}{\mu} + \alpha(\theta - \theta_T), \qquad \Gamma_L(\theta, \sigma) = \frac{\sigma}{\mu} + \alpha(\theta - \theta_T), \quad (10.27)$$

and

$$\left.\begin{aligned}
g_L(\theta, \sigma) &= -\frac{1}{2\mu}\left[\sigma + \mu\alpha(\theta - \theta_T)\right]^2 - \rho c\theta \log(\theta/\theta_T), \\
g_H(\theta, \sigma) &= -\frac{1}{2\mu}\left[\sigma + \mu\alpha(\theta - \theta_T)\right]^2 - \sigma\gamma_T - \rho c\theta \log(\theta/\theta_T) + \rho\lambda_T(\theta - \theta_T)/\theta_T,
\end{aligned}\right\}$$

$$(10.28)$$

where γ_T, λ_T are the transformation strain and latent heat, respectively, at the temperature θ_T. Observe that the two phases have equal Gibbs free energies when the stress vanishes and the temperature is θ_T: it follows that θ_T denotes the transformation temperature. Moreover, since $\Gamma_L(\theta_T, 0) = 0$, the reference configuration is stress free at the transformation temperature. From (10.22) and (10.28) we find that the driving force has the simple expression

$$f = \gamma_T\left[\sigma - \sigma_0(\theta)\right], \qquad (10.29)$$

where the Maxwell stress $\sigma_0(\theta)$ is given by

$$\sigma_0(\theta) = \frac{\rho\lambda_T}{\gamma_T}\left(\frac{\theta - \theta_T}{\theta_T}\right). \qquad (10.30)$$

By using (10.29) and (10.27), one can specialize the system (10.25) to the case of the trilinear material:

$$\left.\begin{aligned}
\dot{s} &= \Phi\left(\gamma_T(\sigma - \sigma_0(\theta)), \theta\right), \\
\delta &= \gamma_T s + \sigma L/\mu + \alpha(\theta - \theta_T)L
\end{aligned}\right\} \qquad (10.31)$$

from which we may eliminate s to get a single master differential equation:

$$\dot{\delta} = \gamma_T\,\Phi\left(\gamma_T(\sigma - \sigma_0(\theta)), \theta\right) + L\dot{\sigma}/\mu + \alpha L\dot{\theta}. \qquad (10.32)$$

We shall discuss initial conditions for (10.32) for the various special loading programs to be described below.

Finally turning to nucleation, one may use (10.29) to express the nucleation thresholds $f_n^{LH}(\theta)$ and $f_n^{HL}(\theta)$ in terms of critical levels of stress $\sigma_n^{LH}(\theta)$ and $\sigma_n^{HL}(\theta)$, thinking of the temperature θ as given, or alternatively as critical levels of temperature $\theta_n^{LH}(\sigma)$ and $\theta_n^{HL}(\sigma)$, thinking of the stress σ as given. For example

$$\sigma_n^{LH}(\theta) = \sigma_0(\theta) + \frac{f_n^{LH}}{\gamma_T} = \frac{\rho\lambda_T}{\gamma_T}\left(\frac{\theta - \theta_T}{\theta_T}\right) + \frac{f_n^{LH}}{\gamma_T}, \qquad (10.33)$$

$$\theta_n^{LH}(\sigma) = \theta_T\left(1 + \frac{\gamma_T\sigma - f_n^{LH}}{\rho\lambda_T}\right), \qquad (10.34)$$

with similar formulas for $\sigma_n^{HL}(\theta)$ and $\theta_n^{HL}(\sigma)$.

Suppose for a moment that the latent heat λ_T is positive. Then according to our earlier discussion, the low-strain phase is austenite and the high-strain phase is martensite. Assume that the bar is in the stress-free reference configuration at the transformation temperature θ_T. With $f_n^{LH} > 0$, the nucleation temperature at zero stress $\theta_n^{LH}(0)$ is necessarily less than θ_T, so that cooling the bar sufficiently at zero stress will cause martensite to nucleate.

10.5 Stress cycles at constant temperature

We now consider a displacement-controlled program of loading in which the elongation rate is prescribed; it is taken to be a positive constant $\dot{\delta}$ during the extensional portion of the loading cycle, after which the bar is shortened at the rate $-\dot{\delta}$. The bar is initially in a stress-free state at a given temperature θ; by $(10.31)_2$ with $s = 0$, $\sigma = 0$, the initial elongation relative to the reference configuration is $\delta_0 = \alpha(\theta - \theta_T)$. First, we wish to determine the relation between the stress $\sigma(t)$ and the elongation $\Delta(t) = \delta(t) - \delta_0$ relative to the initial configuration that results from extending the bar at the given rate $\dot{\delta}$. Until nucleation occurs, the entire bar remains in the low-strain phase, and the desired relation is simply $\sigma = \mu \Delta / L$. At nucleation, which we take to occur at time $t = 0$, the value of Δ is

$$\Delta(0) = \Delta_n^{LH} = \sigma_n^{LH}(\theta)L/\mu. \tag{10.35}$$

Since $\dot{\theta} = 0$ at present, the master equation (10.32) becomes

$$\dot{\delta} = L\dot{\sigma}/\mu + \gamma_T \Phi(\gamma_T(\sigma - \sigma_0(\theta)), \theta). \tag{10.36}$$

Regarding $\Delta = \Delta(0) + \dot{\delta}t$ as the independent variable, one may rewrite (10.36) as

$$\frac{d\sigma}{d\Delta} = \frac{\mu}{L}\left\{1 - \frac{\gamma_T}{\dot{\delta}}\Phi(\gamma_T(\sigma - \sigma_0(\theta)), \theta)\right\} \quad \text{(loading)}. \tag{10.37}$$

The initial condition for (10.37) is

$$\sigma = \sigma_n^{LH} \quad \text{at} \quad \Delta = \Delta_n^{LH}. \tag{10.38}$$

As the bar extends, the moving point in the Δ, σ-plane furnished by (10.37) and (10.38) must always lie in the "flag" described by $\sigma L/\mu \le \Delta \le \gamma_T L + \sigma L/\mu$, $\sigma_m(\theta) \le \sigma \le \sigma_M(\theta)$ for the given temperature θ. If extension continues long enough to carry the phase boundary to the end $x = L$ of the specimen, the bar is fully transformed to the high-strain phase. If extension continues even longer, the point (Δ, σ) will rise along the right flagpole $\sigma = \mu(\Delta/L - \gamma_T)$, in which case the differential equation (10.37) does not apply. When the elongation rate is reversed, the point corresponding to Δ, σ descends along the right flagpole until the reverse transition is nucleated. The shortening history is then determined by the unloading differential equation

$$\frac{d\sigma}{d\Delta} = \frac{\mu}{L}\left\{1 + \frac{\gamma_T}{\dot{\delta}}\Phi(\gamma_T(\sigma - \sigma_0(\theta)), \theta)\right\} \quad \text{(unloading)}, \tag{10.39}$$

along with the initial condition

$$\sigma = \sigma_n^{HL} \quad \text{at} \quad \Delta = \Delta_n^{HL}, \tag{10.40}$$

where $\Delta_n^{HL} = L(\sigma_n^{HL}/\mu + \gamma_T)$ is the nucleation value of Δ on the right flagpole.

If, on the other hand, the sign of the elongation rate were to be reversed *before* the bar has been fully transformed, the shortening process would not involve nucleation of the reverse transition, but rather would take place as in the purely mechanical case, described in detail in Section 3.6.

Indeed, it should be noted that the isothermal, constant-elongation-rate initial-value problem represented by (10.37) and (10.38) and its counterpart for the reverse transition are precise analogs of the corresponding purely mechanical problems posed in Chapter 3 for an elastic bar, with the temperature θ entering the present problem only as a parameter. Thus qualitative features such as initial load rise and load drop that arose in connection with the latter problem occur here as well. Therefore, the new issue of interest in the present setting is how the temperature θ influences the response. To study this issue, it is necessary to make use of a kinetic relation whose θ-dependence is thought to be physically reasonable, at least under suitable circumstances.

To this end, we consider as an example a thermally activated phase transition and make use of a kinetic relation that is a slight modification of the one discussed in Section 8.4.2. In equation (8.26), the velocity of the phase boundary is expressed in terms of the heights of the "energy barriers" confronted by atoms in making either the low-to-high-strain phase transition or the reverse. In present notation, equation (8.26) may be written as

$$\dot{s} = R \left\{ \exp\left(-\frac{b_{LH}}{rk\theta}\right) - \exp\left(-\frac{b_{HL}}{rk\theta}\right) \right\}, \tag{10.41}$$

where b_{LH} and b_{HL} are the respective barrier heights for the forward and reverse transitions, k is Boltzmann's constant, and R and r are absolute constants whose meaning is described in Section 8.4.2. In the model of Section 8.4.2, the barrier heights as given by (8.24) and (8.25) are independent of temperature; for the present trilinear thermoelastic material, this is not the case. One finds instead that

$$b_{LH} = \frac{\gamma_T}{2\,S(\theta)}[\sigma - \sigma_M(\theta)]^2, \qquad b_{HL} = \frac{\gamma_T}{2\,S(\theta)}[\sigma - \sigma_m(\theta)]^2, \tag{10.42}$$

where $S(\theta) = \sigma_M(\theta) - \sigma_m(\theta)$. By using (10.29) to express σ in (10.42) in terms of the driving force f, we may write (10.41) in standard form as

$$\dot{s} = \Phi(f, \theta) = 2R \exp\left\{ -\frac{f^2 + \gamma_T^2\,S^2(\theta)/4}{2\gamma_T rk\theta\,S(\theta))} \right\} \sinh\frac{f}{2rk\theta}. \tag{10.43}$$

Clearly, one has $\Phi(f, \theta)f \geq 0$ as required by the second law; moreover, direct calculation shows that $\Phi(f, \theta)$ increases monotonically with f for $-f_M(\theta) < f < f_M(\theta)$ where $f_M(\theta) = \gamma_T[\sigma_M(\theta) - \sigma_0(\theta)] = \gamma_T[\sigma_M(\theta) - \sigma_m(\theta)]/2 > 0$. Note

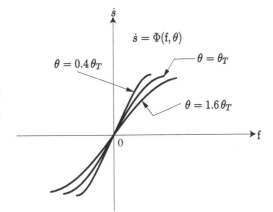

Figure 10.1. Phase boundary velocity as a function of driving force at various temperatures according to the kinetic relation (10.43) of the thermal activation model.

that $\Phi(f, \theta)$ is continuous at $f = 0$, so that $\Phi(0, \theta) = 0$, and therefore the resistance to phase boundary motion vanishes: $f_r^{LH} = f_r^{HL} = 0$.

For a more detailed discussion of thermally activated kinetics for the trilinear thermoelastic material, the reader may consult [1].

We wish to study the loading cycle described by the initial value problem (10.37), (10.38), and (10.43) and its counterpart (10.39), (10.40), and (10.43) for unloading. To describe the numerical details, it is convenient to introduce temporary dimensionless variables as follows:

$$\left.\begin{array}{c} s' = s/L, \quad t' = Rt/L, \quad \sigma' = \sigma/(\mu\gamma_T), \quad f' = f/(\mu\gamma_T^2), \\[2mm] \theta' = (\theta - \theta_T)/\theta_T, \quad \Delta' = \Delta/(L\gamma_T), \quad \dot{\delta}' = \dot{\delta}/(R\gamma_T). \end{array}\right\} \quad (10.44)$$

In addition, we use the following dimensionless material constants:

$$\left.\begin{array}{c} \alpha' = \alpha\theta_T/\gamma_T, \quad M' = M\theta_T/\gamma_T, \quad m' = m\theta_T/\gamma_T, \\[2mm] \theta'_* = (\theta_* - \theta_T)/\theta_T, \quad f'_n = f_n/(\mu\gamma_T^2), \quad k' = kr\theta_T/(\mu\gamma_T^2). \end{array}\right\} \quad (10.45)$$

Except for the dimensionless nucleation parameters f'_n, the dimensionless material constants in (10.45) were assigned a single set of values for the numerical calculations underlying Figures 10.1, 10.2, 10.4 and 10.5:

$$\alpha' = 0.1, \qquad \theta'_* = 1, \qquad M' = 1.15, \qquad m' = 0.85, \qquad k' = 0.1. \quad (10.46)$$

For these choices of the material constants, which are consistent with the restrictions imposed on the trilinear material by equations (9.32) and (9.34), the latent heat λ_T is positive.

Figure 10.1, in which dimensionless phase boundary velocity ds'/dt' is plotted against dimensionless driving force according to the kinetic relation (10.43) for thermal activation, shows the effect of temperature on the kinetic response function Φ. According to the figure, increasing the temperature at a fixed level of driving force *decreases* the phase boundary velocity, although changing the parameters M' and m' can produce the opposite effect. In reporting experiments involving

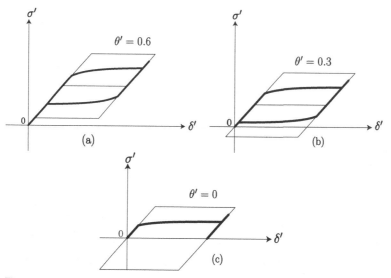

Figure 10.2. Stress-cycling of the bar at three dimensionless temperatures: (a) $\theta' = 0.6$, (b) $\theta' = 0.3$, (c) $\theta' = 0$, each cycle corresponding to the dimensionless elongation rate $\dot{\delta}' = 0.7$. The nucleation values $f_n^{LH'} = 0.14$ and $f_n^{HL'} = -0.14$ were used.

compression of single crystals of a particular CuAlNi alloy, Grujicic, Olson, and Owen [4] show data in their Figure 5 indicating that the velocity of slowly moving phase boundaries *increases* with increasing temperature at each of several values of driving force.

Figure 10.2 presents hysteresis loops corresponding to three different dimensionless temperatures, ranging from $\theta' = 0$, that is, the transformation temperature, to $\theta' = 0.6$, or 1.6 times the transformation temperature. At each temperature, the flag of possible equilibrium states and the corresponding Maxwell line are shown as well. In each of the three cycles, the bar has been extended at the dimensionless elongation rate $\dot{\delta}' = 0.7$ until fully transformed to the high-strain phase; extension of the fully transformed bar continues briefly at this rate. When the elongation rate is reversed, the bar at first shortens in the fully transformed state; at the lowest temperature (Figure 10.2(c)), the bar reaches a stress-free, permanently deformed state before the reverse phase transition can nucleate, and the cycle is stopped. At the two higher temperatures, nucleation of the reverse transition always occurs after sufficient shortening, and the bar is ultimately fully transformed to the low-strain phase, after which it continues to shorten until it returns to the unstressed initial state. As indicated in the figures, increasing the temperature shifts the hysteresis loop in the direction of increasing stress. Figures 10.2(a) and (b) suggest that increasing the temperature does not significantly alter the shape or size of the loop itself.

For purposes of qualitative comparison with Figure 10.2, we reproduce in Figure 10.3 the experimental hysteresis loops for NiTi wires reported by Stoeckel and Yu in [17]; the temperatures in the experiments ranged from 50°C to −40°C.

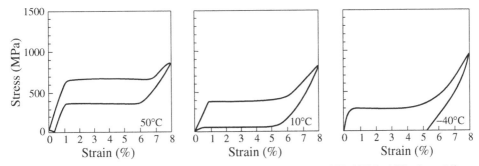

Figure 10.3. Experimental hysteresis loops reported by Stoeckel and Yu [17] for NiTi wires at three different temperatures. This figure was copied with permission from *Wire Journal International* of the Wire Association International. ©WAI Inc.; all rights reserved.

The shapes of the stress–elongation curves shown in Figure 10.2 are *qualitatively* similar to some – but not all – of those determined experimentally for a variety of alloys; see, for example, Grujicic, Olson and Owen [4], Krishnan and Brown [7], Otsuka, Sakamoto and Shimizu [11], Otsuka and Shimizu [12], Pops [13], and Müller and Xu [10]. Permanent deformation of the kind illustrated by Figure 10.2(c) is found to occur at low temperatures but found to be absent at high temperatures in experiments on AgCd [7] and NiTi [12], but some experiments on CuZnSn indicate that permanent deformation may appear at higher temperatures, having been absent at lower ones [13]. That increasing temperature shifts the hysteresis loop in the direction of increasing stress with little effect on the magnitude of the hysteresis itself seems to be consistent with experimental observations on CuZnSi (Pops [13]), CuAlNi (Müller and Xu [10]), and NiTi (Otsuka and Shimizu [12]), but inconsistent with experiments on CuZnSn (Pops [13]). Experiments of Otsuka and Shimizu [12] suggest that the magnitude of the hysteresis may increase or decrease with increasing temperature, depending on the type of martensite involved.

Significant serrations of the kind that are commonly observed in experimental hysteresis loops – see, for example [6] – are not present in those based on the present model and shown in Figure 10.2. Serrations are associated with irregular phase boundary motion and have been modeled by incorporating inhomogeneous behavior in the kinetic relation [2, 14] or by utilizing nonmonotonic kinetic response functions [14].

10.6 Temperature cycles at constant stress

We turn now to a quasistatic process in which the stress is held fixed at the value σ, while the temperature is at first decreased at a given constant rate $\dot{\theta}$ and then increased at the rate $-\dot{\theta}$. We seek the resulting relation between the temperature θ and the elongation Δ from the initial state.

Initially, the specimen is assumed to be in a uniform state at the stress σ and at a temperature greater than the transformation temperature θ_T. During the initial

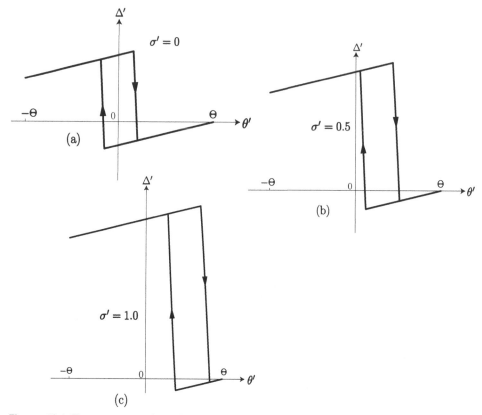

Figure 10.4. Temperature-cycling of the bar at three different dimensionless values of stress $\sigma' = 0, 0.5, 1.0$, each cycle arising from the dimensionless cooling/heating rates $\dot{\theta}' = \mp 0.1$.

period of cooling, the bar remains in the low-strain phase until the temperature reaches the nucleation value $\theta_n^{\mathrm{LH}}(\sigma)$ for the forward transition. Upon nucleation, a phase boundary emerges at $x = 0$ and moves into the bar under the control of the master equation (10.32), which now reduces to a differential equation relating θ and Δ, with the given stress and the cooling rate entering as parameters. The cooling process is continued until somewhat beyond the time at which the bar is fully transformed; heating is then commenced, with the stress remaining at the value σ throughout the experiment. The reverse phase transition ultimately nucleates, the bar eventually returns entirely to the low-strain phase, and the temperature returns to its initial value.

Figure 10.4 results from numerical solutions of the initial-value problems for the master equation that correspond to the thermal loading cycles just described; the dimensionless variables (10.44) and material constants (10.45) were used, with numerical values chosen as they were for Figures 10.1 and 10.2. Three values of the given dimensionless stress are considered, and all three cases involve the same cooling/heating rate. In each case, cooling begins at the point $\theta' = \Theta$, $\Delta' = 0$ in the θ', Δ'-plane and ends when the dimensionless temperature reaches the value $\theta' = -\Theta$; the heating portion of the cycle ultimately returns the bar to the state with

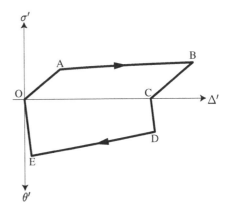

Figure 10.5. The shape-memory effect. The initial temperature is $\theta' = 0$; elongation takes place at the initial temperature at the rate $\dot{\delta}' = 1.0$ along OAB, followed by shortening at the rate $\dot{\delta}' = -1.0$ along BC. Unstressed heating at the rate $\dot{\theta}' = 0.2$ then occurs along CDE, followed by unstressed cooling at the rate $\dot{\theta}' = -0.2$ along EO.

$\theta' = \Theta$, $\Delta' = 0$. The effect of increasing the given stress is to shift the hysteresis loop in the direction of increasing temperature.

Burkhart and Read [3] report results of temperature-cycling experiments on single crystals of indium–thallium at different levels of stress. Because the stresses are compressive, the model under discussion here does not directly apply. It may be noted, however, that the hysteresis loops in Figure 6 of [3] correspond to two different stresses, and as in Figure 10.4 of the present work, the experimental loops are shifted in the direction of increasing temperature upon increase of the magnitude of the stress, and the hysteresis itself does not seem to be affected significantly by the change of stress.

10.7 The shape-memory cycle

The shape-memory effect is exhibited at temperatures for which stress-free permanent deformation can occur during isothermal stress-cycling. We describe the effect in this simple one-dimensional setting with the help of Figure 10.5; note that dimensionless stress σ' is plotted on the upper half of the vertical axis, dimensionless temperature θ' on the lower half, and dimensionless relative elongation Δ' on the horizontal axis. The bar, which is initially stress free in the low-strain phase at, say, the transformation temperature, is stretched at a constant elongation rate at this temperature until it is fully transformed to the high-strain phase (the portion OAB of the stress–elongation curve in Figure 10.5). The elongation rate is then reversed, and the bar is shortened at the initial temperature until it reaches a new stress-free state, now in the high-strain phase (the segment BC of the σ'–Δ' curve in the figure). Next, the stress is maintained at zero while the bar is heated at a constant rate until the reverse transition nucleates (CD in the figure). Heating continues as the low-strain phase grows (DE), eventually enveloping the entire bar at the point E. Finally, the bar is cooled (EO) at zero stress until it recovers its original length at point O.

Figure 10.5 was calculated using the dimensionless quantities introduced in (10.44) and (10.45) with numerical values as in (10.46); as before, the dimensionless nucleation levels of driving force have been chosen as $f_n^{LH'} = 0.14$ and $f_n^{HL'} = -0.14$. The figure is qualitatively similar to the schematic description of the shape-memory effect given by Krishnan et al. [8] in their Figure 11.

10.8 The experiments of Shaw and Kyriakides

Shaw and Kyriakides [15, 16] have carried out experiments on nickel–titanium wires in which the specimen is mechanically cycled at specified elongation rates and temperatures. The temperatures involved in the experiments were in the range of approximately $-20°$C to $100°$C, which includes the temperature range of interest for the austenite–martensite transformations arising in the particular NiTi alloy used. Although the elongation rates in the experiments ranged over four decades, even the highest rates, up to 4×10^{-2} per second, would be considered relatively slow for more conventional metals.

Because the latent heat released or absorbed in phase transitions makes it difficult to maintain a spatially uniform and time-independent temperature in the wire, Shaw and Kyriakides conducted tests in two different, nominally constant-temperature environments: an air-bath, as is conventional, and a liquid bath. In both sets of experiments, up to six small thermocouples were placed along the test section of the wire to measure the local temperature and hence assess its uniformity. To facilitate the measurement of local strains in the specimen, up to four miniature extensometers were positioned along the wire. Detailed results are presented in [15, 16] for tests conducted at $10°$C and at $70°$C.

Those experiments discussed in [15] that were carried out in air exhibited wider variations in specimen temperature than those done in liquid, when both involved the same elongation rate. Temperature variations were found to increase with increasing elongation rate in either environment. In both air and liquid baths, stress cycles at $10°$C exhibited permanent deformation, in qualitative agreement with the cycle predicted by the trilinear thermoelastic model at the lowest of the temperatures shown in Figure 10.2.

As representative of the kinds of data Shaw and Kyriakides were able to obtain, we reproduce in Figure 10.6 a portion of their Figure 7(b) in [15], which shows the results of tests in both air and water at a nominal temperature of $70°$C corresponding to an elongation rate of 4×10^{-4} sec^{-1}. The topmost plot in the figure shows the relation between stress and the *local strain* at an extensometer. The rising portions of the curves represent the rising branches of the underlying rising–falling–rising stress–strain curve; the horizontal portions correspond to the strain jump that occurs as the phase boundary passes the extensometer during the forward (top) and the reverse (bottom) transitions. The local strain at the extensometer and the local temperature at a thermocouple are shown as functions of the overall relative elongation Δ/L in the second (water) and third (air) plots. The time variation in local temperature is larger in air $(6-8°$C$)$ than in water (about $2°$C), with the largest temperature changes occurring when the phase boundary sweeps past the thermocouple.

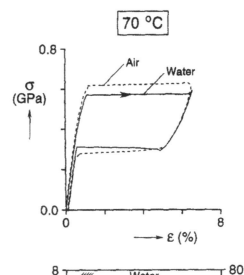

Figure 10.6. The topmost plot represents the local stress–strain responses measured in air and in water at 70°C and a relative elongation rate of 4×10^{-4} sec^{-1}. The local strain and the local temperature as functions of the dimensionless elongation Δ/L are shown in the second and third plots for water and for air, respectively. Reprinted from [15], Figure 7(b), copyright (1995), with permission from Elsevier.

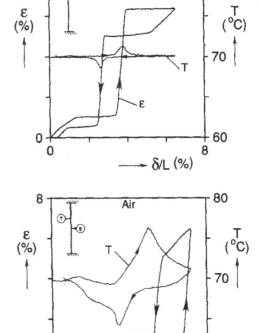

For the purpose of qualitative comparison, we show in Figure 10.7 the counterpart of the experimental strain versus relative elongation graphs of Figure 10.6 as calculated from the trilinear thermoelastic model with thermally activated kinetics. The plot refers to the dimensionless quantities introduced in (10.44) and (10.45) and uses the numerical values (10.46).

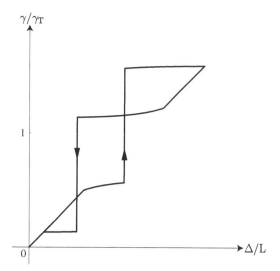

Figure 10.7. Local strain at the particle for which $x/L = 0.3$ as a function of the dimensionless overall elongation Δ' as predicted by the trilinear thermoelastic model with thermally activated kinetics. The loading–unloading cycle is carried out at the dimensionless elongation rates $\dot{\delta}' = \pm 0.7$ and the dimensionless temperature $\theta' = 0.4$. See equations (10.44)–(10.46) for details pertaining to the dimensionless quantities.

In the experiment for which the evolution of the phase transition is described in detail in Sections 4.3 of [16], Shaw and Kyriakides conclude that the forward transition from austenite to martensite nucleates at both ends of the specimen. The experimental setup permits the approximate tracking of moving phase boundaries; the data are consistent with the assumption that the phase boundaries move with constant velocities of 0.25 mm sec^{-1} for the forward transition and 0.32 mm sec^{-1} for the reverse transition from martensite to austenite. These speeds are an order of magnitude faster than the velocity of the moving end grip.

It is reported in [15] that the number of experimentally observed nucleation sites for the austenite-to-martensite transition increases as the elongation rate is increased, with nucleation occurring at a single site only for very slow isothermal experiments. The stresses required to nucleate both forward and reverse transitions are reported as functions of temperature over the range 15–55°C in Figure 7 of [16]. The observed variation of each of these nucleation stresses is consistent with linear dependence on temperature, as predicted by the trilinear thermoelastic model with temperature-*independent* nucleation levels f_n of driving force; see (10.33) in Section 10.4.

10.9 Slow thermomechanical processes

In this section we comment briefly on processes that are slow in the sense that inertial effects can be neglected but are not quasistatic in the preceding sense of being one-parameter families of equilibrium states. When inertial effects are neglected, the field equation (10.4) and jump condition (10.7) associated with linear momentum balance simplify to $\sigma_x = 0$ and $[\![\sigma]\!] = 0$ respectively. The stress field in the bar is therefore uniform. On the other hand the field equation[2] (10.5) associated with the

---

[2] It may be noted that in a quasistatic process in the preceding sense, the energy equation (10.5) is not satisfied since its left-hand side vanishes because of (10.2) but its right-hand side does not vanish in general.

first law of thermodynamics, together with (10.1) and (10.2), yields

$$(K\theta_x)_x = \rho c\theta_t - m\theta\gamma_t, \qquad (10.47)$$

where $c = -\theta\psi_{\theta\theta}$ is the specific heat at constant strain and $m = \rho\psi_{\gamma\theta}$ is the stress–temperature coefficient; cf. (7.23). The related jump condition (10.8) can be written, after using (10.2), (10.9), $\epsilon = \psi + \theta\eta$ and $[\![\theta]\!] = 0$, as

$$[\![K\theta_x]\!] = -(f + \rho\lambda)\dot{s}, \qquad (10.48)$$

where $\lambda = \theta[\![\eta]\!]$ is the latent heat associated with the transition from the phase on the right of the phase boundary to the phase on the left; see (7.14). Observe that the jump condition (10.48) describes a moving point source of heat, the magnitude of which depends on the latent heat λ and the driving force f, and whose current location is $x = s(t)$. The basic problem can therefore be viewed as one involving a moving heat source whose magnitude and propagation speed are not known a priori but must be found by using the kinetic relation.

Suppose that the stress $\sigma(t)$ is prescribed. The strain field can then be expressed in terms of the unknown temperature field by using the constitutive relation (10.17); the partial differential equation (10.47) can thus be written in a form that involves the temperature field as the only unknown. Given suitable initial and boundary conditions, this, together with the jump condition (10.48), can now be solved for $\theta(x, t)$ and $s(t)$, keeping in mind the kinetic relation (10.23) as well as (10.22).

In the uniaxial tension experiments on NiTi carried out by Leo, Shield, and Bruno [9] and Shaw and Kyriakides [15, 16], they observed local heating near the phase boundary when the extension rate was moderately fast. In order to model such experiments, where the effects of thermal diffusion are important but inertial effects are still negligible, one must use the theory outlined above, for example, see Kim and Abeyaratne [5].

REFERENCES

[1] R. Abeyaratne and J.K. Knowles, A continuum model of a thermoelastic material capable of undergoing phase transitions. *Journal of the Mechanics and Physics of Solids*, **41** (1993), pp. 541–71.

[2] K. Bhattacharya, Phase boundary propagation in heterogeneous bodies. *Proceedings of the Royal Society of London* A, **455** (1999), pp. 757–66.

[3] M.W. Burkhart and T.A. Read, Diffusionless phase change in the indium–thallium system. *Journal of Metals, Transactions of the AIME*, **197** (1953), pp. 1516–24.

[4] M. Grujicic, G.B. Olson, and W.S. Owen, Mobility of the $\beta_1 - \gamma'_1$ martensitic interface in CuAlNi. Part I. Experimental measurements. *Metallurgical Transactions*, **16A** (1985), pp. 1723–34.

[5] S.-J. Kim and R. Abeyaratne, On the effect of the heat generated during a stress-induced thermoelastic phase transition. *Continuum Mechanics and Thermodynamics*, **7** (1995), pp. 311–32.

[6] R.V. Krishnan, Stress induced martensitic transformations. *Materials Science Forum*, **3** (1985), pp. 387–98.

[7] R.V. Krishnan and L.C. Brown, Pseudo-elasticity and the strain-memory effect in an Ag-45 at. pct. Cd alloy. *Metallurgical Transactions*, **4** (1973), pp. 423–9.

[8] R.V. Krishnan, L. Delaey, H. Tas, and H. Warlimont, Thermoplasticity, pseudoelasticity and the memory effects associated with martensitic transformations. Part 2. The macroscopic mechanical behavior. *Journal of Materials Science*, **9** (1974), pp. 1536–44.

[9] P.H. Leo, T.W. Shield and O. P. Bruno, Transient heat transfer effects on the pseudoelastic behavior of shape memory wires. *Acta Metallurgica et Materialia*, **41** (1993), pp. 2477–85.

[10] I. Müller and H. Xu, On the pseudo-elastic hysteresis. *Acta Metallurgica et Materialia*, **39** (1991), pp. 263–71.

[11] K. Otsuka, H. Sakamoto, and K. Shimizu, Successive stress-induced martensitic transformations and associated transformation pseudo-elasticity in Cu–Al–Ni alloys. *Acta Metallurgica et Materialia*, **27** (1979), pp. 585–601.

[12] K. Otsuka and K. Shimizu, Pseudo-elasticity and shape-memory effects in alloys. *International Metals Review*, **31** (1986), pp. 93–114.

[13] H. Pops, Stress-induced pseudo-elasticity in ternary Cu–Zn based beta prime phase alloys. *Metallurgical Transactions*, **1** (1970), pp. 251–8.

[14] P. Rosakis and J.K. Knowles, Continuum models for irregular phase boundary motion in solids. *European Journal of Mechanics* A, **18** (1999), pp. 1–16.

[15] J.A. Shaw and S. Kyriakides, Thermomechanical aspects of NiTi. *Journal of the Mechanics and Physics of Solids*, **43** (1995), pp. 1243–81.

[16] J.A. Shaw and S. Kyriakides, On the nucleation and propagation of phase transformation fronts in a NiTi alloy. *Acta Materialia*, **45** (1997), pp. 583–700.

[17] D. Stoeckel and W. Yu, Superelastic Ni–Ti wire. *Wire Journal International*, March, (1991), pp. 45–50.

11 Dynamics of Phase Transitions in Uniaxially Strained Thermoelastic Solids

11.1 Introduction

In Chapter 4 we studied one-dimensional models of dynamic phase transitions in the purely mechanical theory of elastic materials. Our objective here is to extend the ideas of that chapter to the dynamics of two-phase thermoelastic materials. Much of the analysis will be directed to the materials of Mie–Grüneisen type introduced in Chapter 9, with special emphasis on the trilinear thermoelastic material. As in Chapter 4, we shall study an impact-induced phase transition that occurs in compression, rather than in tension; this will require some minor modifications of the details of the constitutive models presented in Chapter 9.

The subject of nonlinear wave propagation in solids has an enormous literature encompassing both experimental and theoretical work. For a sample of background references representing a variety of viewpoints in this field, the reader might consult the classic work on gas dynamics of Courant and Friedrichs [4], the extensive review article of Menikoff and Plohr [7], the discussion by Ahrens [2] of the experimental determination of the "equation of state" of condensed materials, the work of Swegle [9] on phase transitions in materials of geologic interest, and the theory of shock waves in thermoelastic materials presented by Dunn and Fosdick [5].

In the next section, we set out the basic field equations and jump conditions of the dynamical theory of thermoelasticity when the kinematics are those of uniaxial strain and the processes are adiabatic. We specialize these results to materials of Mie–Grüneisen type, focusing on two-phase trilinear materials involving a low-pressure phase (LPP) and a high-pressure phase (HPP). In Section 11.3, we formulate the impact problem for a half-space composed of one of these materials, and we discuss the form of its scale-invariant solutions. For a trilinear material, we first construct solutions corresponding to the absence of a phase transition. These involve a shock wave when the impact is compressive, which is the usual case; for sudden tensile loading, the response involves a fan. When the impact causes a phase transition, we first study solutions in the degenerate special case for which the Grüneisen parameter β vanishes. We then briefly discuss the qualitative departures from these solutions that arise when β is small.

Chen and Lagoudas [3] offer a theory of impact-induced dynamic response in rods for a class of materials for which stress increases monotonically with strain, but whose stiffness at first decreases and then increases as the strain grows, as in the case of rubber; see Figure 3 in [3]. While such a material model predicts two-wave response of the kind observed in impact-induced phase transitions, it does not allow for equilibrium mixtures of phases of the kind that arise in slow loading processes in shape-memory tensile specimens; see [6].

11.2 Uniaxial strain in adiabatic thermoelasticity

11.2.1 Field equations, jump conditions and driving force.
Consider a thermo-elastic body that occupies the slab $0 \le x \le L$, $-\infty < y, z < \infty$ in a reference configuration, which is stress free at the referential temperature; for simplicity, the reference temperature is assumed to be the transformation temperature θ_T of the two-phase material. In uniaxial strain, a particle whose referential location is x, y, z is located at time t at the point $x + u(x, t)$, y, z, where u is the axial displacement. The associated strain and particle velocity are given by $\gamma = u_x$ and $v = u_t$, respectively. Let ρ, $\sigma(x, t)$, and $\epsilon(x, t)$ stand respectively for the referential mass density, the axial component of Piola–Kirchhoff normal stress, and the internal energy per unit mass. In an adiabatic process, the heat flux and external heat supply both vanish identically. Where the fields are smooth, kinematic compatibility and the one-dimensional versions of the balance laws (6.4) and (6.6) of momentum and energy in adiabatic processes require that

$$v_x - \gamma_t = 0, \tag{11.1}$$

$$\sigma_x - \rho v_t = 0, \tag{11.2}$$

$$\sigma \gamma_t - \rho \epsilon_t = 0. \tag{11.3}$$

If there is a discontinuity in strain at time t at the referential location $x = s(t)$, the compatibility and jump conditions (6.10) and (6.13) of momentum and energy when specialized to one-dimensional adiabatic processes yield

$$[\![v]\!] + \dot{s}[\![\gamma]\!] = 0, \tag{11.4}$$

$$[\![\sigma]\!] + \rho \dot{s}[\![v]\!] = 0, \tag{11.5}$$

$$(\rho [\![\epsilon]\!] - \langle \sigma \rangle [\![\gamma]\!]) \dot{s} = 0, \tag{11.6}$$

where $\langle \sigma \rangle = (\sigma^+ + \sigma^-)/2$ denotes the average of the values of σ on either side of the discontinuity.

The local form of the second law of thermodynamics (6.31) requires that for adiabatic processes

$$\eta_t \ge 0 \tag{11.7}$$

where the fields are smooth; according to (6.35), the driving force at a strain discontinuity in an adiabatic process is

$$f = -\rho \langle \theta \rangle [\![\eta]\!], \tag{11.8}$$

so that the entropy inequality

$$f\dot{s} \geq 0 \tag{11.9}$$

requires that $[\![\eta]\!]\,\dot{s} \leq 0$ at a strain discontinuity.

11.2.2 The trilinear Mie–Grüneisen thermoelastic material. It can be readily verified from (9.5) and (9.6) that, for any thermoelastic material, the field equation (11.3) associated with energy balance in an adiabatic process can be written in the equivalent form $\eta_t = 0$. Thus the specific entropy of a particle remains constant during an adiabatic motion (except possibly when a strain discontinuity crosses it). In view of this, it is more convenient in adiabatic processes to describe the constitutive response of a thermoelastic material in terms of the internal energy potential $\hat{\epsilon}(\gamma, \eta)$ rather than the Helmholtz energy potential $\hat{\psi}(\gamma, \theta)$. In uniaxial strain, the thermoelastic constitutive law can be written in the form

$$\sigma = \rho\hat{\epsilon}_\gamma(\gamma, \eta), \qquad \theta = \hat{\epsilon}_\eta(\gamma, \eta), \tag{11.10}$$

where the subscripts γ and η indicate partial derivatives.

It is assumed that the material symmetry of the thermoelastic body under consideration admits uniaxial strain in the x-direction; this will be true, for example, if the material is isotropic in the reference configuration, no matter how the x-direction is chosen. The state of stress will not in general be hydrostatic in uniaxial strain. To calculate the components of normal Piola–Kirchhoff stress transverse to the x-direction, one must know more about the internal energy $\epsilon(\mathbf{F}, \eta)$ for three-dimensional deformation gradient tensors $\mathbf{F}$ than merely its restriction $\hat{\epsilon}(\gamma, \eta)$ to uniaxial strain that appears in (11.10); see Section 7.4. Many treatments of the theory of shock waves in solids assume that the deformation is purely dilatational and the stress tensor is hydrostatic.

Suppose that the slab is composed of a two-phase trilinear Mie–Grüneisen material. Recall from Chapter 9 that a thermoelastic material is said to be of Mie–Grüneisen type if its specific heat at constant strain $c(\gamma, \eta) = \hat{\epsilon}_\eta/\hat{\epsilon}_{\eta\eta}$ is independent of γ, or equivalently if its Grüneisen parameter $\beta(\gamma, \eta) = -\hat{\epsilon}_{\gamma\eta}/\hat{\epsilon}_\eta$ is independent of η. Explicit expressions for the various constitutive relationships for a trilinear Mie–Grüneisen material were given in Section 9.3.1. We shall make use of those expressions in what follows with two formal changes: because we are now considering a phase transition that occurs in compression rather than tension, the transformation strain γ_T is replaced by $-\gamma_T < 0$, and since the reference temperature is here taken to coincide with the transformation temperature we set $\theta_0 = \theta_T$. In particular, the

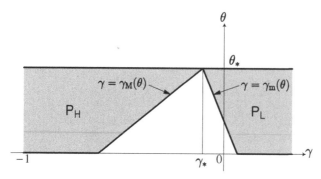

Figure 11.1. The regions in the γ, θ-plane corresponding to the low- and high-pressure phases, LPP and HPP, of the trilinear material associated with (11.11) and (11.12). The regions P_L and P_H are defined by (9.28) and the functions $\gamma_M(\theta)$ and $\gamma_m(\theta)$ are given by (9.29)$_2$.

axial component of stress and the specific entropy are now given by

$$\sigma = \begin{cases} \mu\gamma - \rho c\beta(\theta - \theta_T) & \text{on } P_L, \\ \mu(\gamma + \gamma_T) - \rho c\beta(\theta - \theta_T) & \text{on } P_H, \end{cases} \tag{11.11}$$

$$\eta = \begin{cases} \beta c\gamma + c(1 + \log(\theta/\theta_T)) & \text{on } P_L, \\ \beta c(\gamma + \gamma_T) + c(1 + \log(\theta/\theta_T)) - \lambda_T/\theta_T & \text{on } P_H, \end{cases} \tag{11.12}$$

where the Grüneisen parameter β is defined in terms of the coefficient of thermal expansion α, the modulus μ, the mass density ρ, and the specific heat c by $\beta = \alpha\mu/(\rho c)$, cf. (9.30), (9.35), and (9.9). The regions P_L and P_H of the strain, temperature plane corresponding to the *low-pressure-phase* (LPP) and the *high-pressure-phase* (HPP) are shown in Figure 11.1. The corresponding specific internal energy $\epsilon = \psi + \eta\theta$ is given by

$$\epsilon = \bar{\epsilon}(\gamma, \theta) = \begin{cases} \mu\gamma^2/(2\rho) + c\beta\theta_T\gamma + c\theta & \text{on } P_L, \\ \mu(\gamma + \gamma_T)^2/(2\rho) + c\beta\theta_T(\gamma + \gamma_T) + c\theta & \text{on } P_H. \end{cases} \tag{11.13}$$

In order to determine the potential $\hat{\epsilon}(\gamma, \eta)$ one would solve (11.12) for θ as a function of γ and η and substitute the result into (11.13).

A moving strain discontinuity at $x = s(t)$ is a shock wave if the states on both sides belong to the same material phase, so that the points $(\gamma^+(t), \theta^+(t))$ and $(\gamma^-(t), \theta^-(t))$ are either both in the region P_L or both in P_H. If the points are in different regions, the discontinuity is a phase boundary. For the trilinear material, explicit formulas for the driving force at a strain discontinuity can be readily found by using (11.12) to specialize (11.8). This leads, for a shock wave, to

$$f = -\rho c\langle\theta\rangle \left(\beta(\gamma^+ - \gamma^-) + \log(\theta^+/\theta^-)\right) \tag{11.14}$$

and, for a phase boundary with LPP on the right, to

$$f = -\rho c\langle\theta\rangle \left\{\beta(\gamma^+ - \gamma^- - \gamma_T) + \log(\theta^+/\theta^-) + \lambda_T/(c\theta_T)\right\}. \tag{11.15}$$

11.3 The impact problem

11.3.1 Formulation: Scale-invariant solutions.
We now reduce the field equations (11.1), (11.2), and (11.3) to three differential equations for strain γ, particle velocity v, and temperature θ. Using (11.11) in (11.2) and (11.12) in (11.3) after noting that the latter reduces to $\eta_t = 0$ for any thermoelastic material, one is led to

$$v_x - \gamma_t = 0, \tag{11.16}$$

$$a^2 \gamma_x - c\beta\theta_x - v_t = 0, \tag{11.17}$$

$$\beta\gamma_t + \theta_t/\theta = 0, \tag{11.18}$$

for *either phase* of the trilinear material. Here we have written

$$a = \sqrt{\mu/\rho} \tag{11.19}$$

for the isothermal wave speed of linearized theory.

We assume now that the slab is semi-infinite in the x-direction, and that initially, it is at rest in its reference configuration at the transformation temperature. The material is therefore in LPP initially. At time $t = 0$, a prescribed constant velocity v_0 is given to the particles at the boundary $x = 0$ of the half-space and maintained for all subsequent times. Thus we seek weak solutions of the differential equations (11.16)–(11.18) in the quadrant $x > 0$, $t > 0$ satisfying the following initial and boundary conditions

$$\gamma(x, 0) = 0, \qquad v(x, 0) = 0, \qquad \theta(x, 0) = \theta_T \quad \text{for } x > 0, \tag{11.20}$$

$$v(0, t) = v_0 \quad \text{for } t > 0, \tag{11.21}$$

where, unless otherwise stated, the given impactor velocity v_0 is assumed to be positive, causing compression. We have chosen the initial temperature to be the transformation temperature θ_T to simplify the formulas. The more general case where $\theta(x, 0) \neq \theta_T$ is treated in [1].

At a shock wave in either phase of the trilinear material, the jump conditions (11.4)–(11.6) specialize to

$$[\![v]\!] + \dot{s}[\![\gamma]\!] = 0, \tag{11.22}$$

$$a^2[\![\gamma]\!] - c\beta[\![\theta]\!] + \dot{s}[\![v]\!] = 0, \tag{11.23}$$

$$\{[\![\theta]\!] + \beta\langle\theta\rangle[\![\gamma]\!]\}\dot{s} = 0. \tag{11.24}$$

By (11.14), the entropy inequality (11.9) at a shock wave in either phase becomes

$$\{\beta[\![\gamma]\!] + \log(\theta^+/\theta^-)\}\dot{s} \leq 0. \tag{11.25}$$

At a phase boundary with LPP on the right, the jump conditions (11.4)–(11.6) take the forms

$$[\![v]\!] + \dot{s}[\![\gamma]\!] = 0, \tag{11.26}$$

$$a^2([\![\gamma]\!] - \gamma_T) - c\beta[\![\theta]\!] + \dot{s}[\![v]\!] = 0, \tag{11.27}$$

$$\{-\gamma_T a^2(\langle\gamma\rangle + \gamma_T/2) + c\beta\langle\theta\rangle [\![\gamma]\!] + c[\![\theta]\!] + \lambda_T - c\beta\gamma_T\theta_T\}\dot{s} = 0, \tag{11.28}$$

and the entropy inequality (11.9) with f given by (11.15) is

$$\{c\beta([\![\gamma]\!] - \gamma_T) + c\log(\theta^+/\theta^-) + \lambda_T/\theta_T\}\dot{s} \le 0. \tag{11.29}$$

The boundary–initial value problem posed above is invariant under the scale change $x \to kx$, $t \to kt$, for any nonzero constant k. We seek solutions with this same invariance; in such solutions, γ, v, and θ must be functions of $\zeta = x/t$ only. For scale-invariant solutions, the system (11.16)–(11.18) reduces to

$$\left.\begin{array}{r} \zeta\,\gamma'(\zeta) + v'(\zeta) = 0, \\[4pt] a^2\gamma'(\zeta) - c\beta\theta'(\zeta) + \zeta\,v'(\zeta) = 0, \\[4pt] \beta\gamma'(\zeta) + \theta'(\zeta)/\theta(\zeta) = 0. \end{array}\right\} \tag{11.30}$$

Viewed as a linear homogeneous system for γ', θ', and v', (11.30) always has the trivial solution corresponding to a uniform state in which the strain, temperature, and particle velocity are constants. There is a nontrivial solution if and only if the determinant of the system vanishes. This condition turns out to be

$$\theta(\zeta) = \frac{1}{c\beta^2}(\zeta^2 - a^2). \tag{11.31}$$

When (11.31) holds, γ and v are easily shown from (11.30) to have the structure

$$\gamma(\zeta) = \frac{1}{\beta}\log(\zeta^2 - a^2) + A, \quad v(\zeta) = \frac{2}{\beta}\left\{\zeta + \frac{a}{2}\log\left(\frac{\zeta - a}{\zeta + a}\right)\right\} + B, \tag{11.32}$$

where A and B are arbitrary constants. Thus for a scale-invariant solution of the impact problem, the first quadrant of the x, t-plane may be partitioned into sectors of the form $\zeta_1 t < x < \zeta_2 t$, on each of which γ, v, and θ are either constants or given by the fan (11.31) and (11.32).

11.3.2 Solutions with no phase transition. For sufficiently small impactor velocities v_0, one expects the target to remain in LPP. We accordingly seek first a scale-invariant solution of the impact problem in which a shock wave moves into the unstrained material at rest at the temperature θ_T. Thus we write

$$\gamma = \begin{cases} \gamma^-, & 0 < x < \dot{s}t, \\[4pt] 0, & x > \dot{s}t, \end{cases} \quad v = \begin{cases} v_0, & 0 < x < \dot{s}t, \\[4pt] 0, & x > \dot{s}t, \end{cases} \quad \theta = \begin{cases} \theta^-, & 0 < x < \dot{s}t, \\[4pt] \theta_T, & x > \dot{s}t, \end{cases} \tag{11.33}$$

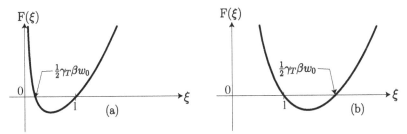

Figure 11.2. The function $F(\xi)$ that determines the dimensionless shock wave speed ξ from (11.36). (a) $w_0 < 2/(\gamma_T \beta)$ and (b) $w_0 > 2/(\gamma_T \beta)$.

and seek to determine the three unknowns γ^-, θ^-, and $\dot{s}$ from the three jump conditions (11.22)–(11.24) with the entropy inequality (11.25) enforced. It is useful to note first that the three homogeneous equations (11.22)–(11.24) for the jumps $[\![\gamma]\!]$, $[\![v]\!]$, and $[\![\theta]\!]$ have nontrivial solutions if and only if

$$\dot{s} = a\sqrt{1 + (c\beta^2/a^2)\langle\theta\rangle}, \tag{11.34}$$

so that necessarily the shock wave speed $\dot{s}$ exceeds the isothermal wave speed a. Eliminating θ^- and γ^- from the three jump conditions provides an equation for the shock wave speed $\dot{s}$. To describe the results, it is convenient to introduce dimensionless quantities by writing

$$T = \frac{c}{\gamma_T^2 a^2}\theta, \qquad w_0 = \frac{v_0}{\gamma_T a}, \qquad \xi = \frac{\dot{s}}{a}. \tag{11.35}$$

Then one finds that the dimensionless shock wave speed ξ is given by the roots of the equation

$$F(\xi) = \gamma_T^2 \beta^2 T_T \quad \text{where} \quad F(\xi) = \frac{(\xi^2 - 1)(\xi - \frac{1}{2}\gamma_T\beta\, w_0)}{\xi}; \tag{11.36}$$

as depicted in Figure 11.2, the function $F(\xi)$ in $(11.36)_2$ is convex for $\xi > 0$. Once ξ has been found from (11.36), the strain γ^- and the dimensionless temperature T^- behind the shock are given by

$$\gamma^- = -\gamma_T \frac{w_0}{\xi}, \qquad T^- = \frac{\xi + \frac{1}{2}\gamma_T\beta\, w_0}{\xi - \frac{1}{2}\gamma_T\beta\, w_0} T_T. \tag{11.37}$$

Note that, since the dimensionless impactor velocity w_0 is positive, (11.37) shows that the strain γ^- is compressive, and the temperature behind the shock is higher than that in front.

For small dimensionless impactor velocity w_0, (11.36) gives $\xi \sim \sqrt{1 + 2T_T}$, which, in dimensionless terms, is the wavespeed of linearized isentropic thermoelastic waves in uniaxial strain.

Graphs of the relation (11.36) between dimensionless phase boundary speed $\xi = \dot{s}/a$ and dimensionless impactor velocity $w_0 = v_0/(\gamma_T a)$ for three different dimensionless initial temperatures $T_T = c\theta_T/(\gamma_T^2 a^2)$ are shown in Figure 11.3.

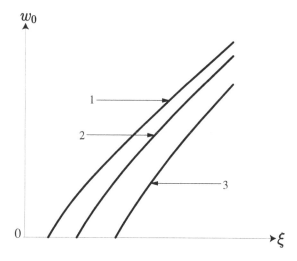

Figure 11.3. The relation between the dimensionless shock wave speed $\xi = \dot{s}/a$ and the dimensionless impactor velocity $w_0 = v_0/(\gamma_T a)$ according to (11.36) at three different dimensionless initial temperatures $T_T = c\theta_T/(\gamma_T^2 a^2)$. Curve 1: $T_T = 12$, Curve 2: $T_T = 31.5$, Curve 3: $T_T = 71.5$.

For future purposes, we record the form taken by the shock-wave solution described in (11.33) and (11.35)–(11.37) in the special case $\beta = \alpha = 0$:

$$\xi = \dot{s}/a = 1, \qquad \gamma^- = -\gamma_T w_0 = -v_0/a, \quad T^- = T_T, \quad (\beta = 0). \quad (11.38)$$

To assure that the solution to the impact problem just constructed involves only the LPP of the material, one must require that the point (γ^-, θ^-) lies in the region P_L in the strain–temperature plane, so that

$$\gamma^- > \gamma_m(\theta^-) = \gamma_* + m(\theta^- - \theta_*), \quad 0 < \theta^- < \theta_*; \quad (11.39)$$

see $(9.29)_2$ and Figure 11.1. Ultimately, this requirement imposes an upper limit on the impactor velocity, which depends on the many material parameters entering the theory. The form taken by this upper limit in the special case $\beta = \alpha = 0$ is

$$v_0/a < \frac{\gamma_T}{2} - \frac{1}{2}(M - m)(\theta_* - \theta_T). \quad (11.40)$$

When the impactor velocity v_0 is negative, producing tension, there is still a formal solution consisting of a single shock wave, but it can be shown to violate the entropy inequality. The proper solution in this case can be constructed from the fan (11.31) and (11.32); such a construction turns out to be possible *only* when $v_0 < 0$.

11.3.3 Solutions with a phase transition. When the impact problem results in a phase transition, one typically confronts a propagating phase boundary, at which the jump conditions (11.26)–(11.28) and the entropy inequality (11.29) must hold, and a shock wave, at which (11.22)–(11.24) and (11.25) must be satisfied. Analytical construction of a piecewise constant, closed-form solution fulfilling these requirements as well as the initial and boundary conditions is not practical, but most of the qualitative features of such a solution can be appreciated by looking at the special case in which the Grüneisen parameter $\beta = \alpha a^2/c$ is small, which we study by perturbing about the special case $\beta = \alpha = 0$.

The case $\beta = \alpha = 0$. When $\beta = 0$, the system (11.16)–(11.18) reduces to

$$a^2\gamma_x - v_t = 0, \qquad v_x - \gamma_t = 0, \tag{11.41}$$

$$\theta_t = 0, \tag{11.42}$$

where the fields are smooth. At a shock wave, the jump conditions (11.22) and (11.23) for compatibility and momentum with $\beta = 0$ are

$$[\![v]\!] + \dot{s}[\![\gamma]\!] = 0, \qquad a^2[\![\gamma]\!] + \dot{s}[\![v]\!] = 0, \tag{11.43}$$

while the energy jump condition (11.24) becomes simply

$$[\![\theta]\!]\dot{s} = 0; \tag{11.44}$$

the entropy inequality (11.25) at the shock reduces to

$$\left\{\log(\theta^+/\theta^-)\right\}\dot{s} \le 0. \tag{11.45}$$

At a phase boundary, the jump conditions (11.26) and (11.27) with $\beta = 0$ go to

$$[\![v]\!] + \dot{s}[\![\gamma]\!] = 0, \qquad a^2[\![\gamma]\!] + \dot{s}[\![v]\!] - \gamma_T a^2 = 0, \tag{11.46}$$

for compatibility and momentum, and the jump condition (11.28) for energy balance becomes

$$\left\{c[\![\theta]\!] - \gamma_T a^2[\langle\gamma\rangle + \gamma_T/2] + \lambda_T\right\}\dot{s} = 0. \tag{11.47}$$

The entropy inequality (11.29) at the phase boundary with $\beta = 0$ is

$$\left\{(\lambda_T/\theta_T) + c\log(\theta^+/\theta^-)\right\}\dot{s} \le 0. \tag{11.48}$$

In this special case $\beta = 0$, the field equations and jump conditions that arise from momentum balance and compatibility - (11.41), (11.43), and (11.46) - are precisely those for a trilinear *elastic* material in the purely mechanical setting of Chapter 4, with shock waves traveling at the speed $a = \sqrt{\mu/\rho}$. The consequences (11.42) and (11.44) of energy balance, along with the entropy inequality (11.45) at a shock, show that, when $\beta = 0$, temperature enters only trivially unless a phase boundary is present. If there is a phase boundary, the associated energy balance (11.47) *does* involve coupling of thermal and mechanical quantities.

When a phase transition is present, we seek a solution in which the leading disturbance is a shock wave propagating into undisturbed material, followed by a phase boundary. Thus we take

$$\gamma, v, \theta = \begin{cases} \gamma^-, \ v_0, \ \theta^- & \text{for } 0 \le x < \dot{s}t, \\ \gamma^+, \ v^+, \ \theta_T & \text{for } \dot{s} < x < at, \\ 0, \ 0, \ \theta_T & \text{for } x > at, \end{cases} \tag{11.49}$$

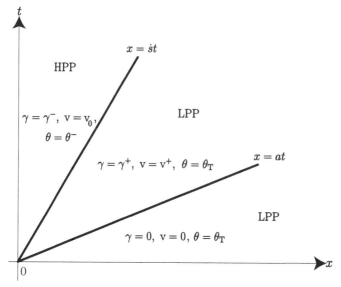

Figure 11.4. The x, t-plane for a solution of the impact problem in which a phase transition occurs (special case $\beta = 0$).

where $\dot{s}$ is the phase boundary velocity; see Figure 11.4. Upon enforcing the jump conditions (11.43), (11.46), and (11.47), one finds

$$\gamma^- = -\frac{v_0}{a} - \gamma_T \frac{a}{a + \dot{s}}, \tag{11.50}$$

$$\theta^- = \theta_T + \frac{\lambda_T}{c} + \frac{a\gamma_T}{c} v_0 + \frac{a^2\gamma_T^2}{2c} \frac{\dot{s}^2 - 2a\dot{s}}{a^2 - \dot{s}^2}, \tag{11.51}$$

$$\gamma^+ = -\frac{v_0}{a} + \gamma_T \frac{a\dot{s}}{a^2 - \dot{s}^2}, \tag{11.52}$$

$$v^+ = v_0 - \gamma_T \frac{a^2\dot{s}}{a^2 - \dot{s}^2}, \tag{11.53}$$

where $\dot{s}$ is the as yet undetermined phase boundary velocity.

The entropy inequality (11.29) becomes

$$\frac{\lambda_T}{c\theta_T} - \log\left\{1 + \frac{\lambda_T}{c\theta_T} + \frac{\gamma_T a^2}{2c\theta_T}\left[2\frac{v_0}{a} + \gamma_T \frac{\dot{s}^2 - 2a\dot{s}}{a^2 - \dot{s}^2}\right]\right\} \le 0. \tag{11.54}$$

Using (11.50)–(11.53) in (11.49) furnishes a one-parameter family of phase-changing solutions to the impact problem in the case $\beta = 0$. The as yet undetermined phase boundary velocity $\dot{s}$ and the given impactor velocity v_0 must not only satisfy the entropy inequality (11.54), but they are subject to the inequalities that follow from the phase segregation requirements and the limits on the range of θ^-:

$$\gamma^+ > \gamma_m(\theta_T) = \gamma_* + m(\theta_T - \theta_*),$$

$$-1 < \gamma^- < \gamma_M(\theta^-) = \gamma_* + M(\theta^- - \theta_*), \tag{11.55}$$

$$0 < \theta^- < \theta_*.$$

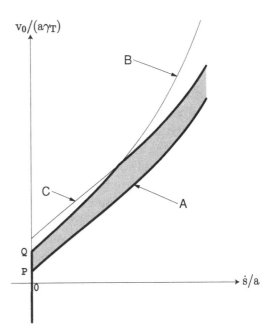

Figure 11.5. Admissible phase boundary velocity, impactor velocity pairs correspond to points in the shaded region in the $\dot{s}/a$, $v_0/(a\gamma_T)$-plane. Curve A consists of points for which $f = 0$, or $[\![\eta]\!] = 0$; on curve B, $\gamma^+ = \gamma_m(\theta_T)$; and on curve C, $\theta^- = \theta_*$. Curves B and C intersect. The four dimensionless parameters in (11.56) have been given the values $N = 1, n = -3, T_T = 0.01, T_* = 0.2$.

When (11.50)–(11.53) are used in (11.55), the resulting inequalities restrict the dimensionless velocities $v_0/(\gamma_T a)$ and $\dot{s}/a$ and involve dimensionless parameters given by

$$N = \gamma_T a^2 M/c, \quad n = \gamma_T a^2 m/c, \quad T_T = c\theta_T/(\gamma_T^2 a^2), \quad T_* = c\theta_*/(\gamma_T^2 a^2).$$
(11.56)

The restrictions may be described in a plane in which $\dot{s}/a$ and $v_0/(\gamma_T a)$ are the coordinates, as indicated, for example, in Figure 11.5. In constructing the curves in the figure, the four dimensionless parameters in (11.56) have been given the values $N = 1, n = -3, T_T = 0.01$, and $T_* = 0.2$. The shaded region in the figure represents, in their dimensionless versions, the pairs $\dot{s}, v_0$ that fulfill the three requirements $\gamma^+ > \gamma_m(\theta_T), \theta^- < \theta_*$ and the entropy inequality $f\dot{s} \geq 0$. The limitations imposed by the additional inequalities $\gamma^- > \gamma_M(\theta^-)$ and $\theta^- > 0$ require admissible points to lie above two curves, not shown in the figure, that themselves lie everywhere below curve A ($f = 0$) that forms the lower boundary of the shaded region; therefore these additional inequalities impose no relevant restrictions. The curve (not shown) in the plane of Figure 11.5 on which $\gamma^- = -1$ actually cuts off the shaded region at very large values of v_0, corresponding to extreme compression, a regime in which the trilinear model is likely to be wholly unrealistic. A similar caveat applies to the limitation imposed by the portion of the upper boundary of the shaded region that lies above the point of intersection of curves B and C. This limitation arises when the temperature θ^- behind the phase boundary approaches the critical temperature θ_*, another regime in which the trilinear model is not to be trusted.

Let us return for a moment to the solution (11.38) to the impact problem *without* a phase transition in the special case $\beta = \alpha = 0$. The restriction (11.40) on the impactor velocity v_0 assuring that the state behind the shock wave is in the LPP corresponds precisely to the portion of the vertical axis lying below the point Q and shown bold in Figure 11.5.

As the given impactor velocity v_0 is increased from zero, the shock-wave solution with the specimen remaining in the LPP is the only possible response to impact until v_0 reaches the value corresponding to the point P in Figure 11.5. For values of v_0 between P and Q, the no-phase-change solution and a one-parameter family of solutions with a phase boundary are both possible. As in the purely mechanical case, the nucleation criterion chooses between these two types of solutions. When it determines that a phase transition must occur, the kinetic relation then selects a particular member of the one-parameter family of solutions with a phase transition. Based on the general discussion in Section 8.2, one form for a kinetic relation is

$$\dot{s} = \Phi(f/\langle\theta\rangle), \tag{11.57}$$

where f is the driving force acting on the phase boundary, $\langle\theta\rangle$ is the average of the temperatures on either side, and the kinetic response function Φ depends on the material. To assure that the second law of thermodynamics is satisfied at the phase boundary, one requires that Φ satisfy $\Phi(z)z \geq 0$ for all real z. If $\Phi(z)$ increases monotonically with z, (11.57) can be inverted to give the kinetic relation in the form

$$f = \langle\theta\rangle\phi(\dot{s}). \tag{11.58}$$

By inserting the representations (11.50)–(11.52) for γ^-, γ^+, and θ^- into (11.15), one can find the driving force on the phase boundary. Using this and (11.51) in (11.58) yields the relation between the given impactor velocity v_0 and the unknown phase boundary velocity $\dot{s}$ needed to complete the solution of the impact problem when a phase transition occurs:

$$\frac{v_0}{a} = G(\dot{s}) + \frac{c\theta_T}{\gamma_T a^2}\exp(l)\{\exp[\phi(\dot{s})/(\rho c)] - 1\} \quad (\beta = 0), \tag{11.59}$$

where $l = \lambda_T/(c\theta_T)$ is the dimensionless latent heat, and

$$G(\dot{s}) = \frac{c\theta_T}{\gamma_T a^2}(e^l - 1 - l) + \frac{\gamma_T}{2}(2a\dot{s} - \dot{s}^2)/(a^2 - \dot{s}^2). \tag{11.60}$$

The curve A that forms the lower boundary of the shaded admissible region in Figure 11.5 is described by the equation $v_0/(a\gamma_T) = G(\dot{s})$. It can be shown that the right side of (11.59) increases monotonically with $\dot{s}$; one must of course restrict $\phi(\dot{s})$ so that the curve represented by (11.59) lies in the shaded region of the figure. Specifying a value for the impactor velocity v_0 will then lead to a uniquely determined phase boundary speed $\dot{s}$.

Some numerical details such as the dependence on impactor velocity of the final temperature and final density attained after the passage of the phase boundary are given in [1] for the special case of zero latent heat and for the linear kinetic relation $\dot{s} = M f / \langle \theta \rangle$, where the mobility M is constant. The case of vanishing latent heat arises in the twinning of crystals, when the two material "phases" are actually variants of a single phase.

The case $\beta \neq 0$. To get some indication of the modifications to the results of the preceding subsection that arise when β does not vanish, one can assume that the Grüneisen parameter $\beta = \alpha a^2 / c$ is small and solve the impact problem approximately by perturbation with respect to β. Such an assumption is not quantitatively appropriate for solids, as one sees by scanning Table 5.1 on p. 133 of [8], where referential values of the modified Grüneisen parameter are given in the rightmost column. Though we shall not carry out the small-g perturbation process here, we shall describe the main qualitative changes that take place when the assumption $\beta = 0$ is replaced by $\beta \ll 1$; for further details, see [1].

The principal new feature that appears when one considers the case $\beta \neq 0$ with $\beta \ll 1$ is the possibility that the precursor to the phase boundary may be *either* a shock wave, as in the case $\beta = 0$, *or* a narrow fan, which cannot occur when $\beta = 0$. The approximate form for small β of the condition that requires a fan, rather than a shock, to precede the phase boundary is determined in [1], and it turns out to permit a certain range of positive, or compressive, values of the impactor velocity v_0 under some circumstances. As we saw earlier, in the absence of a phase transition, a fan can arise only for "tensile impact" ($v_0 < 0$), whatever the value of β.

REFERENCES

[1] R. Abeyaratne and J.K. Knowles, Impact-induced phase transitions in thermoelastic solids. *Philosophical Transactions of the Royal Society of London* A, **355** (1997), pp. 843–67.

[2] T.J. Ahrens, Equation of state, in *High Pressure Shock Compression of Solids*, edited by J.R. Asay and M. Shahinpoor. Springer-Verlag, New York, 1993, pp. 75–114.

[3] Y.-C. Chen and D.C. Lagoudas, Impact induced phase transformation in shape memory alloys. *Journal of the Mechanics and Physics of Solids*, **48** (2000), pp. 275–300.

[4] R. Courant and K.O. Friedrichs, *Supersonic Flow and Shock Waves*. Interscience, New York, 1948.

[5] J.E. Dunn and R.L. Fosdick, Steady, structured shock waves. Part 1. Thermoelastic materials. *Archive for Rational Mechanics and Analysis*, **104** (1988), pp. 295–365.

[6] J.K. Knowles, Impact-induced tensile waves in a rubberlike material. *SIAM Journal on Applied Mathematics*, **62** (2002), pp. 1153–75.

[7] R. Menikoff and B.J. Plohr, The Riemann problem for flow of real materials. *Reviews of Modern Physics*, **61** (1989), pp. 75–130.

[8] M.A. Meyers, *Dynamic Behavior of Materials*. Wiley-Interscience, New York, 1994.

[9] J.W. Swegle, Irreversible phase transitions and wave propagation in silicate geologic materials. *Journal of Applied Physics*, **68** (1990), pp. 1563–79.

Part V Higher Dimensional Problems

12 Statics: Geometric Compatibility

12.1 Preliminaries

In this chapter we consider the simplest possible static problem in three dimensions that involves a phase boundary, that is, a continuous, piecewise homogeneous, two-phase deformation. Despite its simplicity, this problem serves to illustrate a number of phenomena that do not occur in one dimension. In particular, we will demonstrate through a series of examples that the existence of such deformations in a crystalline solid depends sensitively on the lattice parameters. In some materials such deformations only exist for special values of the lattice parameters and in others, they cannot exist at all. The reader may consult the book by Bhattacharya [4] for a complete discussion of the material in this Chapter.

Consider a thermoelastic body that occupies all of $\mathbb{R}^3$, and suppose that it is subjected to a piecewise homogeneous deformation

$$\mathbf{y} = \begin{cases} \mathbf{F}^+\mathbf{x} & \text{for } \mathbf{x} \cdot \mathbf{n} \geq 0, \\ \mathbf{F}^-\mathbf{x} & \text{for } \mathbf{x} \cdot \mathbf{n} \leq 0, \end{cases} \tag{12.1}$$

where $\mathbf{F}^\pm$ are constant tensors belonging to two distinct phases or variants. The planar interface $S = \{\mathbf{x} : \mathbf{x} \cdot \mathbf{n} = 0\}$ corresponds to the referential image of a phase or twin boundary and the unit vector $\mathbf{n}$ is normal to this interface.

Suppose that the temperature θ of the body is uniform. Suppose further that the deformation gradient tensors $\mathbf{F}^\pm$ lie at the bottoms of two energy wells of the potential $\psi(\mathbf{F}, \theta)$. Then the corresponding stresses vanish: $\sigma^\pm = \rho\psi_{\mathbf{F}}(\mathbf{F}^\pm, \theta) = \mathbf{0}$. Finally, suppose that the body is at the transformation temperature, $\theta = \theta_T$, so that $\psi(\mathbf{F}^+, \theta_T) = \psi(\mathbf{F}^-, \theta_T)$. Then, the driving force on the phase boundary also vanishes: $f = \rho(\psi(\mathbf{F}^+, \theta_T) - \psi(\mathbf{F}^-, \theta_T)) - \sigma^\pm \cdot (\mathbf{F}^+ - \mathbf{F}^-) = 0$. Consequently the body is in thermal, mechanical, and phase equilibria.

All that remains is to enforce geometric compatibility across the phase boundary. The deformation gradient tensor suffers a jump discontinuity across S but suppose that the deformation itself is continuous. This requires that

$$\mathbf{F}^+\boldsymbol{\ell} = \mathbf{F}^-\boldsymbol{\ell} \tag{12.2}$$

for all vectors ℓ that are tangent to the phase boundary, which in turn holds if and only if there is a nonzero vector $\mathbf{a}$ such that

$$\mathbf{F}^+ = \mathbf{F}^- + \mathbf{a} \otimes \mathbf{n}, \quad \mathbf{a} \neq \mathbf{0}; \tag{12.3}$$

see (6.8) and (6.9). Thus $\mathbf{F}^+$ and $\mathbf{F}^-$ differ by a rank-1 tensor.

We now propose to study the following two issues: given two distinct symmetric positive definite tensors $\mathbf{U}^+$ and $\mathbf{U}^-$,

(i) under what conditions on $\mathbf{U}^\pm$ does the compatibility condition (12.3) hold for some vectors $\mathbf{a}$, $\mathbf{n}$, and some rotation tensors $\mathbf{R}^\pm$ with $\mathbf{F}^\pm = \mathbf{R}^\pm \mathbf{U}^\pm$? and

(ii) when (12.3) does hold, find $\mathbf{a}$, $\mathbf{n}$, and $\mathbf{R}^\pm$.

It is convenient to first rephrase these issues in forms that do not involve the rotation tensors. Observe that if $\mathbf{a}$, $\mathbf{n}$, $\mathbf{R}^+$, $\mathbf{R}^-$ satisfy (12.3) for some $\mathbf{U}^+$ and $\mathbf{U}^-$, then so do $\mathbf{Q}\mathbf{a}$, $\mathbf{n}$, $\mathbf{Q}\mathbf{R}^+$, $\mathbf{Q}\mathbf{R}^-$ for any rotation $\mathbf{Q}$. Thus with no loss of generality we can pick one of the rotation tensors $\mathbf{R}^\pm$ arbitrarily and so we take $\mathbf{R}^- = \mathbf{I}$. Next, let $\mathbf{D}$, $\mathbf{b}$, and $\mathbf{m}$ be defined by

$$\mathbf{D} = \mathbf{F}^+(\mathbf{F}^-)^{-1}, \qquad \mathbf{b} = |(\mathbf{F}^-)^{-T}\mathbf{n}|\mathbf{a}, \qquad \mathbf{m} = (\mathbf{F}^-)^{-T}\mathbf{n}/|(\mathbf{F}^-)^{-T}\mathbf{n}|. \tag{12.4}$$

On using (12.4) we can write (12.3) as $\mathbf{D} = \mathbf{I} + \mathbf{b} \otimes \mathbf{m}$. Finally, set $\mathbf{C} = \mathbf{D}^T\mathbf{D}$ so that $\mathbf{C}$ is symmetric and positive definite and note from $(12.4)_1$ that it can be written as

$$\mathbf{C} = (\mathbf{U}^-)^{-1}(\mathbf{U}^+)^2(\mathbf{U}^-)^{-1}. \tag{12.5}$$

The geometric compatibility equation (12.3) now yields

$$\mathbf{C} = (\mathbf{I} + \mathbf{m} \otimes \mathbf{b})(\mathbf{I} + \mathbf{b} \otimes \mathbf{m}). \tag{12.6}$$

The two issues presented above can now be posed in the following equivalent form: given two distinct stretch tensors $\mathbf{U}^+$ and $\mathbf{U}^-$, define the symmetric positive definite tensor $\mathbf{C}$ by (12.5) and then (i) find conditions on $\mathbf{C}$ under which (12.6) holds for some nonnull vectors $\mathbf{b}$ and $\mathbf{m}$ and (ii) when these conditions hold, find $\mathbf{b}$ and $\mathbf{m}$. It can be readily verified that, once $\mathbf{b}$ and $\mathbf{m}$ have been found, $\mathbf{R}^+$ determined from $\mathbf{D} = \mathbf{F}^+(\mathbf{F}^-)^{-1} = \mathbf{R}^+\mathbf{U}^+(\mathbf{U}^-)^{-1} = \mathbf{I} + \mathbf{b} \otimes \mathbf{m}$, together with $\mathbf{F}^+ = \mathbf{R}^+\mathbf{U}^+$, $\mathbf{F}^- = \mathbf{R}^-\mathbf{U}^-$, and $\mathbf{R}^- = \mathbf{I}$, allow $\mathbf{n}$ and $\mathbf{a}$ to be determined from $(12.4)_{3,2}$ and that the resulting set $\{\mathbf{F}^+, \mathbf{F}^-, \mathbf{a}, \mathbf{n}\}$ satisfies the original compatibility condition (12.3). The following proposition answers the rephrased questions just posed:

Proposition: Ball and James [2]. Let $0 < \lambda_1 \leq \lambda_2 \leq \lambda_3$ be the ordered eigenvalues of $\mathbf{C}\,(\neq \mathbf{I})$ and let $\mathbf{e}_1, \mathbf{e}_2, \mathbf{e}_3$ be a set of corresponding orthonormal eigenvectors. Then a necessary and sufficient condition for there to exist vectors $\mathbf{b}\,(\neq \mathbf{0})$ and $\mathbf{m}$ satisfying (12.6) is that

$$\lambda_1 \leq 1, \qquad \lambda_2 = 1, \qquad \lambda_3 \geq 1. \tag{12.7}$$

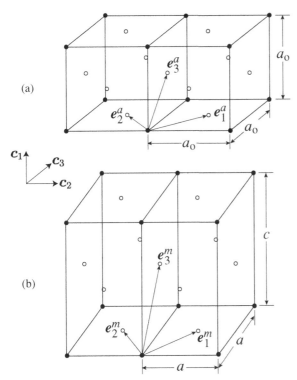

Figure 12.1. (a) Face-centered cubic lattice with lattice parameters $a_o \times a_o \times a_o$ and lattice vectors $\{e_1^a, e_2^a, e_3^a\}$ and (b) face-centered tetragonal lattice with lattice parameters $a \times a \times c$ and lattice vectors $\{e_1^m, e_2^m, e_3^m\}$. The unit vectors $\{c_1, c_2, c_3\}$ are associated with the cubic directions. Solely for clarity, the atoms at the vertices are depicted by (small) filled circles while those at the centers of the faces are shown by open circles.

When (12.7) holds, there are two pairs of vectors $\mathbf{b}$, $\mathbf{m}$ for which (12.6) holds and they are given by

$$
\left.
\begin{aligned}
\mathbf{b} &= d \left\{ \sqrt{\frac{\lambda_3(1 - \lambda_1)}{\lambda_3 - \lambda_1}} \mathbf{e}_1 + \kappa \sqrt{\frac{\lambda_1(\lambda_3 - 1)}{\lambda_3 - \lambda_1}} \, \mathbf{e}_3 \right\}, \\
\mathbf{m} &= \frac{1}{d} \frac{\sqrt{\lambda_3} - \sqrt{\lambda_1}}{\sqrt{\lambda_3 - \lambda_1}} \left\{ -\sqrt{1 - \lambda_1} \, \mathbf{e}_1 + \kappa \sqrt{\lambda_3 - 1} \, \mathbf{e}_3 \right\},
\end{aligned}
\right\}
\tag{12.8}
$$

where d is chosen to make $\mathbf{m}$ a unit vector and $\kappa = \pm 1$. For a proof of these results the reader is referred to[1] [2].

We now apply this result to a number of examples, all of which except the final one concern a material possessing a *cubic austenite phase* and a *tetragonal martensite phase* (as considered previously in Section 5.5). Suppose that the cubic lattice has lattice parameters $a_o \times a_o \times a_o$, and let $\{\mathbf{c}_1, \mathbf{c}_2, \mathbf{c}_3\}$ be a set of orthonormal vectors in the cubic directions; see Figure 12.1. The tetragonal lattice, with lattice

---

[1] The analysis of this proposition in [2] is framed within the larger setting of sequences of energy minimizing deformations. For a purely geometric analysis, see Bowles and Mackenzie [5], and Wechsler Lieberman and Read, [11].

parameters $a \times a \times c$, is obtained by stretching the parent lattice equally in two cubic directions by a stretch ratio η_1 and in the third direction by a ratio η_3:

$$\eta_1 = a/a_o, \qquad \eta_3 = c/a_o. \tag{12.9}$$

There are three distinct ways in which to carry out this transformation corresponding to three variants of martensite characterized by stretch tensors $\mathbf{U}_1$, $\mathbf{U}_2$, $\mathbf{U}_3$:

$$\mathbf{U}_k = \eta_1 \mathbf{I} + [\eta_3 - \eta_1] \mathbf{c}_k \otimes \mathbf{c}_k, \quad k = 1, 2, 3. \tag{12.10}$$

The components of these stretch tensors in the cubic basis are

$$\underline{U}_1 = \begin{pmatrix} \eta_3 & 0 & 0 \\ 0 & \eta_1 & 0 \\ 0 & 0 & \eta_1 \end{pmatrix}, \qquad \underline{U}_2 = \begin{pmatrix} \eta_1 & 0 & 0 \\ 0 & \eta_3 & 0 \\ 0 & 0 & \eta_1 \end{pmatrix}, \qquad \underline{U}_3 = \begin{pmatrix} \eta_1 & 0 & 0 \\ 0 & \eta_1 & 0 \\ 0 & 0 & \eta_3 \end{pmatrix}. \tag{12.11}$$

Since the reference configuration has been taken to coincide with unstressed austenite, it corresponds to the deformation gradient tensor $\mathbf{I}$.

In all of the examples below, with the exception of the last one, we shall not explicitly write out the solutions $\mathbf{b}$, $\mathbf{m}$, $\mathbf{a}$, $\mathbf{n}$, $\mathbf{R}^+$. We will focus instead on the implications of the condition (12.7) for piecewise homogeneous deformations. In particular, the examples will illustrate how (12.7) restricts the possibilities, in some cases disallowing such deformations, in others allowing them always, and in yet others, allowing them under special circumstances.

12.2 Examples

Example 1: *Two variants of martensite.* First consider a piecewise homogeneous deformation involving two of the three martensitic variants, say $\mathbf{U}_1$ and $\mathbf{U}_2$. Now $\mathbf{F}^+ = \mathbf{R}^+ \mathbf{U}_2$ and $\mathbf{F}^- = \mathbf{U}_1$ and so from (12.5), $\mathbf{C} = \mathbf{U}_1^{-1} \mathbf{U}_2^2 \mathbf{U}_1^{-1}$. The components of $\mathbf{C}$ are found using (12.11) to be

$$\underline{C} = \begin{pmatrix} \eta_1^2/\eta_3^2 & 0 & 0 \\ 0 & \eta_3^2/\eta_1^2 & 0 \\ 0 & 0 & 1 \end{pmatrix}, \tag{12.12}$$

whence the ordered eigenvalues of $\mathbf{C}$ are η_1^2/η_3^2, 1, η_3^2/η_1^2 (if $\eta_1 < \eta_3$), or η_3^2/η_1^2, 1, η_1^2/η_3^2 (if $\eta_3 < \eta_1$). Thus the requirements (12.7) are automatically satisfied thereby guaranteeing the existence of a piecewise homogeneous deformation involving two variants. A deformation involving two martensitic variants is called a *twin* or *twinning deformation*; the interface between the variants is a *twin boundary*.

Example 2: *Austenite and one variant of martensite.* Next consider the possibility of a piecewise homogeneous deformation involving austenite on one side of the interface and one variant of martensite, say $\mathbf{U}_1$, on the other side. Thus we take $\mathbf{F}^+ = \mathbf{R}^+ \mathbf{U}_1$ and $\mathbf{F}^- = \mathbf{I}$ and find from (12.5) that $\mathbf{C} = \mathbf{U}_1^2$. Thus from (12.11) we have

$$\underline{C} = \begin{pmatrix} \eta_3^2 & 0 & 0 \\ 0 & \eta_1^2 & 0 \\ 0 & 0 & \eta_1^2 \end{pmatrix}. \tag{12.13}$$

It follows that the requirement (12.7) holds if and only if

$$\eta_1 = 1, \qquad \eta_3 \neq 1. \tag{12.14}$$

The condition $\eta_1 = 1$ is a restriction on the lattice and implies that the lattice parameters must obey $a = a_o$. Such an austenite $\rightarrow$ martensite transformation involves a change in only *one* lattice dimension, the cubic and tetragonal lattice dimensions being $a_o \times a_o \times a_o$ and $a_o \times a_o \times c$ respectively. Thus unless the material is special in this way, no sharp interfaces separating cubic austenite from one variant of tetragonal martensite will be observed.

This is consistent with the observation that in most martensitic materials one does not observe a simple piecewise homogeneous deformation during phase transformation involving two homogeneously deformed phases separated by a planar phase boundary; for example, see Bhattacharya [4]. Usually, the crystal-lographic structure on one side of a phase boundary is more complicated than this. It should be noted however that there do exist some exceptional materials that exhibit simple piecewise homogeneous deformations during phase transfor-mation, such as the titanium–tantalum alloys at certain compositions as studied by Bywater and Christian [6]. It is known that varying the alloy composition changes the lattice parameters, so that presumably, it may sometimes be possible to pick a composition for which the lattice parameters satisfy the requisite spe-cial relationships, in which event that alloy will exhibit piecewise homogeneous deformations of the form studied in this example.

The result of Example 2 suggests that we examine austenite/martensite de-formations that have a more complicated structure than that in Example 2. A complete treatment of the next two examples is beyond the scope of the present discussion since they must be viewed from the point of view of sequences of de-formations with increasingly finer microstructure; see Ball and James [2]. Here we restrict our attention to a discussion of the macroscopic kinematics.

Example 3: *Austenite and twinned martensite.* (Ball and James [2]) Let us now consider the deformation described schematically in Figure 12.2 in which there is an "interface" which separates austenite from martensite. The martensitic half-space $\mathbf{m} \cdot \mathbf{x} < 0$ involves bands of martensite, which alternate between variants -1 and -2. Thus the martensitic region is twinned, and Example 1 describes its kinematics. The deformation gradient tensors in the martensitic layers are

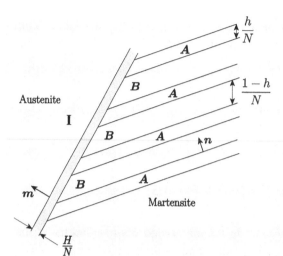

Figure 12.2. An interface separating austenite from twinned martensite. The martensitic variants are characterized by the deformation gradient tensors **A**, **B** and have volume fractions h and $1 - h$. The thickness of the shaded diffuse interface tends to zero as the fineness of the twin bands increases, that is, $N \to \infty$.

$\mathbf{A} = \mathbf{Q}_2\mathbf{U}_2$ and $\mathbf{B} = \mathbf{Q}_1\mathbf{U}_1$ and geometric compatibility across the martensite/martensite interfaces requires that

$$\mathbf{A} - \mathbf{B} = \mathbf{a}' \otimes \mathbf{n}. \tag{12.15}$$

The widths of the martensitic layers are proportional to h/N and $(1 - h)/N$ where the volume fraction h is as yet unknown and N is a positive integer. We have seen in Example 2 that neither variant of martensite (characterized by **A** or **B**) is compatible with austenite (characterized by **I**). Therefore, at best, we can only join austenite to martensite through an interpolation layer, corresponding to a diffuse interface of width H/N as depicted by the gray strip in Figure 12.2. Consider now the sequence of deformations indexed by N. As $N \to \infty$ the diffuse interface collapses to a plane. It is shown in [2] that if the local deformation gradient field is to remain bounded during this limiting process, it is necessary and sufficient that the deformation gradient **I** in the austenite region be compatible with the *average* deformation gradient $h\mathbf{A} + (1 - h)\mathbf{B}$ in the martensite region. In other words, as $N \to \infty$, the austenite region "sees" not the individual martensitic variants but the average of the two. Compatibility between the austenite and the averaged martensite across the austenite/martensite interface requires that

$$[h\mathbf{A} + (1 - h)\mathbf{B}] - \mathbf{I} = \mathbf{b} \otimes \mathbf{m}. \tag{12.16}$$

Given $\mathbf{U}_1$ and $\mathbf{U}_2$, we wish to solve the pair of equations (12.15) and (12.16) with $\mathbf{A} = \mathbf{Q}_2\mathbf{U}_2$, $\mathbf{B} = \mathbf{Q}_1\mathbf{U}_1$ and determine $\mathbf{a}'$, $\mathbf{n}$, $\mathbf{b}$, $\mathbf{m}$, $\mathbf{Q}_1$, and $\mathbf{Q}_2$.

In order to put this into the standard form, set $\mathbf{R} = \mathbf{Q}_1^T\mathbf{Q}_2$, $\bar{\mathbf{R}} = \mathbf{Q}_1$, and $\mathbf{a} = \mathbf{Q}_1^T\mathbf{a}'$ so that the two compatibility equations can be rewritten as

$$\mathbf{R}\mathbf{U}_2 - \mathbf{U}_1 = \mathbf{a} \otimes \mathbf{n}, \qquad \bar{\mathbf{R}}(\mathbf{U}_1 + h\mathbf{a} \otimes \mathbf{n}) = \mathbf{I} + \mathbf{b} \otimes \mathbf{m}. \tag{12.17}$$

The first of these is the twinning equation, which we examined previously in Example 1 and found to be solvable under all conditions. The second equation leads to (12.6) provided we set

$$\mathbf{C} = (\mathbf{U}_1 + h\mathbf{n} \otimes \mathbf{a})(\mathbf{U}_1 + h\mathbf{a} \otimes \mathbf{n}). \tag{12.18}$$

If (12.6) is to be solvable, $\mathbf{C}$ must satisfy the conditions (12.7); in particular, it is necessary that one of the eigenvalues of $\mathbf{C}$ be unity, and therefore that $\det(\mathbf{C} - \mathbf{I}) = 0$. An explicit expression for $\mathbf{C}$ can be obtained by using (12.18) and substituting into it (12.11) for $\mathbf{U}_1$, and the solution of Example 1 for $\mathbf{a}$ and $\mathbf{n}$. This leads to

$$\mathbf{C} = \frac{[h(\eta_3^2 - \eta_1^2) - \eta_3^2]^2 + \eta_1^2\eta_3^2}{\eta_1^2 + \eta_3^2} \, \mathbf{c}_1 \otimes \mathbf{c}_1 + \frac{[h(\eta_3^2 - \eta_1^2) + \eta_1^2]^2 + \eta_1^2\eta_3^2}{\eta_1^2 + \eta_3^2} \, \mathbf{c}_2 \otimes \mathbf{c}_2$$

$$\pm \frac{(\eta_3^2 - \eta_1^2)^2}{\eta_3^2 + \eta_1^2} \, h(1 - h) \, (\mathbf{c}_1 \otimes \mathbf{c}_2 + \mathbf{c}_2 \otimes \mathbf{c}_1) + \eta_1^2 \, \mathbf{c}_3 \otimes \mathbf{c}_3, \tag{12.19}$$

which expresses $\mathbf{C}$ in terms of the lattice parameters η_1, η_3, the orthonormal cubic basis vectors $\{\mathbf{c}_1, \mathbf{c}_2, \mathbf{c}_3\}$, and the unknown volume fraction h; the $\pm$ here arises from the fact that the twinning equation $(12.17)_1$ has two solutions. Setting $\det(\mathbf{C} - \mathbf{I}) = 0$ now leads to a quadratic equation for the volume fraction h whose roots are

$$h = \frac{1}{2} \left(1 \pm \sqrt{1 + \frac{2 \left(\eta_1^2 - 1\right) \left(\eta_3^2 - 1\right) \left(\eta_3^2 + \eta_1^2\right)}{\left(\eta_3^2 - \eta_1^2\right)^2}} \right). \tag{12.20}$$

Note that the sum of these two roots is unity. When (12.20) holds, the eigenvalues of $\mathbf{C}$ are found to be $1, \eta_1^2, \eta_1^2\eta_3^2$. The necessary and sufficient condition (12.7), together with the inequality that ensures that (12.20) yields real roots in the interval $[0, 1]$, can now be shown to be equivalent to

$$\text{either} \quad \eta_1 < 1 < \eta_3 \quad \text{and} \quad \eta_1^{-2} + \eta_3^{-2} \leq 2,$$

$$\text{or} \quad \eta_3 < 1 < \eta_1 \quad \text{and} \quad \eta_1^2 + \eta_3^2 \leq 2. \tag{12.21}$$

Thus in summary, an austenite-twinned martensite deformation of the assumed form (Figure 12.2) exists provided that the lattice parameters obey (12.21) and the variant volume fraction is given by (12.20). Table 7.1 of Bhattacharya [4] compares these results with experimental observations for a number of alloys possessing cubic austenite and tetragonal martensite; the agreement is seen to be good.

Example 4: *Wedge of twinned martensite in austenite.* (Bhattacharya [3]) As a final example consider the deformation illustrated in Figure 12.3. Here one has a wedge of martensite surrounded by austenite. The wedge involves a mid-rib, on one side of which there is a fine mixture of two martensitic variants characterized by deformation gradients $\mathbf{A}$ and $\mathbf{B}$ and volume fractions h_1 and $1 - h_1$; on the other side there is a second fine mixture of martensitic variants characterized by the deformation gradients $\mathbf{C}, \mathbf{D}$ and volume fractions h_2 and $1 - h_2$.

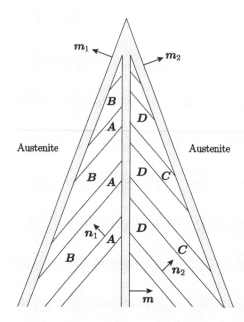

Figure 12.3. Wedge of twinned martensite surrounded by austenite. The microstructure involves five types of interfaces: two sets of twin boundaries (separating different variants of martensite, **A**/**B** and **C**/**D**), two austenite/martensite interfaces, and the mid-rib of the wedge, which separates twinned martensite **A**/**B** from twinned martensite **C**/**D**.

There are five types of interfaces to be considered: two sets of twin boundaries, the two austenite–martensite boundaries, and the mid-rib. Enforcing geometric compatibility across these interfaces leads to the following five conditions:

martensite/martensite : $\mathbf{A} - \mathbf{B} = \mathbf{a}_1 \otimes \mathbf{n}_1$,

martensite/martensite : $\mathbf{C} - \mathbf{D} = \mathbf{a}_2 \otimes \mathbf{n}_2$,

austenite/twinned martensite : $[h_1\mathbf{A} + (1 - h_1)\mathbf{B}] - \mathbf{I} = \mathbf{b}_1 \otimes \mathbf{m}_1$,

austenite/twinned martensite : $[h_2\mathbf{C} + (1 - h_2)\mathbf{D}] - \mathbf{I} = \mathbf{b}_2 \otimes \mathbf{m}_2$,

twinned martensite/twinned martensite :

$$[h_1\mathbf{A} + (1 - h_1)\mathbf{B}] \ - [h_2\mathbf{C} \ + (1 - h_2)\mathbf{D}] = \mathbf{b} \otimes \mathbf{m}. \quad (12.22)$$

The first two of these are twinning equations and we know from Example 1 that they are solvable without the need for any restrictions. The next two equations are precisely of the type analyzed in Example 3. We know that these are solvable provided the volume fractions h_1 and h_2 of the variants are suitably chosen (and the lattice parameters satisfy some mild inequalities). This completely determines all of the free parameters in the problem. The solvability of the final equation therefore imposes a relationship between the lattice parameters, which Bhattacharya [3] has shown to be

$$\eta_1^2 = \frac{\left(1 - \eta_3^2\right)^2 + 4\eta_3^2\left(1 + \eta_3^2\right)}{\left(1 - \eta_3^2\right)^2 + 8\eta_3^4}. \quad (12.23)$$

Thus we conclude that a deformation of the assumed form involving a wedge of martensite can only exist in special materials where the lattice parameters satisfy the restriction (12.23).

Bhattacharya [3, 4] has conducted an extensive survey of the literature and shown that a variety of cubic-tetragonal martensitic materials do in fact satisfy this restriction and that these materials exhibit martensitic wedges; there are also materials that do not satisfy (12.23) and they do not exhibit wedge shaped martensitic domains.

Example 5: *Austenite and one variant of martensite*; see [1, 7–10]. In Example 2 we observed that the homogeneously deformed cubic phase can be separated by a planar interface from a single variant of homogeneously deformed *tetragonal* phase only if the lattice parameters are related in a special way. We now examine this issue for cubic and *monoclinic* phases. A material that exhibits such phases is a CuAlNi shape memory alloy where the austenite is cubic and the β_1'-martensite phase is monoclinic. We will find, again, that a two-phase piecewise homogeneous deformation is possible only if the lattice geometry is suitably restricted (by (12.28)). In Chapter 13 we shall study the transformation between these particular phases in an impact experiment.

In order to describe a monoclinic martensitic phase it is convenient to consider the deformation to it from the cubic austenite in two stages. Let $\{c_1, c_2, c_3\}$ be a set of orthonormal vectors in the cubic directions of the austenite lattice (Figure 12.4(i)). First, stretch the material by a stretch ratio η_2 in, say, the c_2 direction and by η_3 and η_1 along the diagonals of the face normal to c_2; see Figure 12.4. Thus if $\{d_1, d_2, d_3\}$ are orthonormal vectors obtained by rotating the cubic basis through $\pi/4$ about c_2, so that

$$\mathbf{d}_1 = \frac{1}{\sqrt{2}}\mathbf{c}_1 - \frac{1}{\sqrt{2}}\mathbf{c}_3, \qquad \mathbf{d}_2 = \mathbf{c}_2, \qquad \mathbf{d}_3 = \frac{1}{\sqrt{2}}\mathbf{c}_1 + \frac{1}{\sqrt{2}}\mathbf{c}_3, \qquad (12.24)$$

then the stretch tensor $\mathbf{U}$ characterizing this deformation is

$$\mathbf{U} = \eta_3 \mathbf{d}_1 \otimes \mathbf{d}_1 + \eta_2 \mathbf{d}_2 \otimes \mathbf{d}_2 + \eta_1 \mathbf{d}_3 \otimes \mathbf{d}_3. \qquad (12.25)$$

The cubic lattice has now transformed into an orthorhombic lattice (Figure 12.4(ii)). Second, subject the orthorhombic lattice to a simple shear on a plane normal to $\mathbf{d}_1$, in the direction $\mathbf{d}_3$, by an amount k. The deformation gradient tensor $\mathbf{K}$ associated with this simple shear is

$$\mathbf{K} = \mathbf{I} + k\,\mathbf{d}_3 \otimes \mathbf{d}_1, \qquad (12.26)$$

and the orthorhombic lattice has now been mapped into a monoclinic lattice (Figure 12.4(iii)). The gradient of the composite homogeneous deformation, from the cubic lattice to the monoclinic lattice, is $\mathbf{KU}$. This describes one of the monoclinic variants. Since there are six ways in which to carry out the first part of the deformation, and for each of them, there are two ways in which to carry out the second deformation, there are a total of twelve monoclinic martensitic variants. Note that the geometry of the monoclinic lattice is described by the four parameters η_1, η_2, η_3, and k.

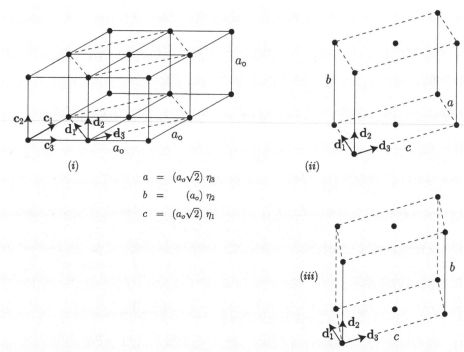

$$a = (a_o\sqrt{2})\,\eta_3$$
$$b = (a_o)\,\eta_2$$
$$c = (a_o\sqrt{2})\,\eta_1$$

Figure 12.4. The cubic lattice (i), transformed into an orthorhombic lattice (ii), which in turn is transformed into a monoclinic lattice (iii).

Suppose that the cubic austenite and the monoclinic martensite coexist with a planar interface between them. Then we know from our preliminary discussion that there must exist a rotation tensor $\mathbf{R}$ and vectors $\mathbf{b}$ and $\mathbf{m}$ such that

$$\mathbf{RKU} = \mathbf{I} + \mathbf{b} \otimes \mathbf{m}. \qquad (12.27)$$

Given the stretch tensor $\mathbf{U}$ and the shear tensor $\mathbf{K}$, one can find a rotation tensor $\mathbf{R}$ and vectors $\mathbf{b}$ and $\mathbf{m}$ if and only if the eigenvalues $\lambda_1, \lambda_2, \lambda_3$ of $(\mathbf{KU})^T\mathbf{KU}$ obey (12.7). It can be readily shown that the condition $\lambda_2 = 1$ in (12.7) is satisfied only if the amount of shear k in (12.26) is related to the stretches η_i in (12.25) by

$$k = \sqrt{\frac{(\eta_1^2 - 1)\,(\eta_3^2 - 1)}{\eta_3^2}}. \qquad (12.28)$$

Thus if a piecewise homogeneous two-phase deformation is to be possible, the condition (12.28), which is a restriction on the geometry of the monoclinic lattice, must hold. If in addition to (12.28) the transformation stretches are such that

$$\text{either} \quad \eta_2 \leq 1, \quad \eta_1\eta_3 \geq 1, \quad \text{or} \quad \eta_2 \geq 1, \quad \eta_1\eta_3 \leq 1, \qquad (12.29)$$

one can show that all requirements in (12.7) then hold. Thus, a necessary and sufficient condition for homogeneously deformed cubic austenite and monoclinic martensite to coexist across a planar interface is that (12.28) and (12.29) hold.

In Chapter 13 we shall study the impact-induced transformation between cubic and monoclinic phases in a particular alloy satisfying the restrictions (12.28) and (12.29). The following formulas will be useful in that analysis: when (12.28) holds, the eigenvalues of $(\mathbf{KU})^T\mathbf{KU}$ are

$$\eta_2^2, \quad 1, \quad \eta_1^2\eta_3^2, \tag{12.30}$$

with corresponding orthonormal eigenvectors $\{\mathbf{p}_1, \mathbf{p}_2, \mathbf{p}_3\}$

$$\left.\begin{aligned} \mathbf{p}_1 &= \mathbf{d}_2, \\[2mm] \mathbf{p}_2 &= -\sqrt{\frac{\eta_1^2-1}{\eta_1^2\eta_3^2-1}}\,\mathbf{d}_1 + \eta_1\sqrt{\frac{\eta_3^2-1}{\eta_1^2\eta_3^2-1}}\,\mathbf{d}_3, \\[2mm] \mathbf{p}_3 &= \eta_1\sqrt{\frac{\eta_3^2-1}{\eta_1^2\eta_3^2-1}}\,\mathbf{d}_1 + \sqrt{\frac{\eta_1^2-1}{\eta_1^2\eta_3^2-1}}\,\mathbf{d}_3. \end{aligned}\right\} \tag{12.31}$$

According to (12.8), (12.30), and (12.31), the unit normal $\mathbf{m}$ to the austenite–martensite interface is given by

$$\mathbf{m} = \sqrt{\frac{1-\eta_2^2}{\eta_1^2\eta_3^2-\eta_2^2}}\,\mathbf{p}_1 - \kappa\sqrt{\frac{\eta_1^2\eta_3^2-1}{\eta_1^2\eta_3^2-\eta_2^2}}\,\mathbf{p}_3, \tag{12.32}$$

and the vector $\mathbf{b}$ is given by

$$\mathbf{b} = -(\eta_1\eta_3-\eta_2)\left[\sqrt{\frac{\eta_1^2\eta_3^2(1-\eta_2^2)}{\eta_1^2\eta_3^2-\eta_2^2}}\,\mathbf{p}_1 + \kappa\sqrt{\frac{\eta_2^2(\eta_1^2\eta_3^2-1)}{\eta_1^2\eta_3^2-\eta_2^2}}\,\mathbf{p}_3\right], \tag{12.33}$$

where $\kappa = \pm 1$.

REFERENCES

[1] R. Abeyaratne and J.K. Knowles, On the kinetics of an austenite→martensite phase transformation induced by impact in a Cu–Al–Ni shape memory alloy. *Acta Materialia*, **45** (1997), pp. 1671–83.

[2] J.M. Ball and R.D. James, Fine phase mixtures as minimizers of energy. *Archive for Rational Mechanics and Analysis*, **100** (1987), pp. 13–52.

[3] K. Bhattacharya, Wedge-like microstructure in martensite. *Acta Metallurgica Materialia*, **10** (1991), pp. 2431–44.

[4] K. Bhattacharya, *Microstructure of Martensite: Why it Forms and How it Gives Rise to the Shape Memory Effect*. Oxford University Press, New York, 2003.

[5] J.S. Bowles and J.K. Mackenzie, The crystallography of martensite transformations. *Acta Metallurgica Materialia*, **2** (1954), pp. 129–47.

[6] K.A. Bywater and J.W. Christian, Martensitic transformations in titanium–tantalum alloys. *Philosophical Magazine*, **25** (1972), pp. 1249–73.

[7] J.C. Escobar, *Plate Impact Induced Phase Transformations in Cu–Al–Ni Single Crystals*. Ph.D. dissertation, Brown University, Providence, RI, 1995.

[8] J.C. Escobar and R.J. Clifton, On pressure–shear plate impact for studying the kinetics of stress-induced phase transformations. *Journal of Materials Science and Engineering*, **A170** (1993), pp. 125–42.

[9] J.C. Escobar and R.J. Clifton, Pressure–shear impact-induced phase transitions in Cu-14.4 Al-4.19 Ni single crystals. *SPIE*, **2427** (1995), pp. 186–97.

[10] R.D. James and K.F. Hane, Martensitic transformations and shape memory materials. *Acta Materialia*, **48** (2000), pp. 197–222.

[11] M.S. Wechsler, D.S. Lieberman, and T.A. Read, On the theory of the formation of martensite. *Transactions of the AIME Journal of Metals*, **197** (1953), pp. 1503–15.

13 Dynamics: Impact-Induced Transition in a CuAlNi Single Crystal

13.1 Introduction

In this chapter we determine the kinetic law for a particular material directly from experimental data. The material is a single crystal of a CuAlNi shape memory alloy and the experiments were conducted by Escobar and Clifton [3–5] using impact loading to induce a transformation from cubic austenite to monoclinic β_1'-martensite.

In their experiments Escobar and Clifton used thin platelike specimens, which were subjected to impact loading on one face with the resulting particle velocity measured on the opposite face. The orientation of the lattice relative to the specimen was carefully arranged so as to lead to the propagation of a phase boundary that was parallel to the faces of the specimen. In addition, the magnitude and direction of impact, which involved both normal and shear components, were chosen so as to induce the desired transformation.

The material tested was a single crystal of Cu–14.44Al–4.19Ni (wt%). This shape memory alloy involves an austenite phase that is cubic, one martensite phase (called γ_1') that is orthorhombic, and a second martensite phase (called β_1') that is monoclinic; see Example 5 in Section 12.2. Depending on the stress and temperature, a different one of these phases is energetically favorable. A schematic phase diagram for this material showing the preferred phases at different temperatures and stresses was shown previously in Figure 5.1. In the experiments described in the present chapter the material transforms from austenite to β_1'-martensite. In Chapter 14 we will be concerned with some quasistatic experiments at very low stress on (essentially) this same alloy, this time involving γ_1'-martensite.

In order to determine the kinetic law we have to determine the driving force and the propagation speed of the phase boundary. However the impact generates three elastic waves, which propagate well ahead of the phase boundary. Since the measurement of particle velocity during the experiment could only be made for some small time after impact, the measurements in fact ended well *before* the phase boundary reached the back face of the specimen. Thus the measurements do not allow one to directly determine the arrival time of the phase boundary at the back face of the specimen or the particle velocity after its passage. Both of these quantities are needed in order to determine the phase boundary propagation speed and the driving force. Consequently an analysis needs to be carried out in order to infer these quantities from the measured data.

In an unstressed state, austenite and martensite are associated with the bottoms of their respective energy wells. In our analysis we assume that during the motion, the deformation gradient tensor at any point in the specimen does not depart significantly from this state. The deformation gradient tensor *is* permitted to jump from one energy well to the other, though once it is associated with a particular energy well it is assumed to be near its bottom. This allows us to linearize the field equations for each phase, though the mathematical problem is still nonlinear since the two domains occupied by the two phases are separated by a moving interface at which the appropriate jump conditions hold. This perturbation analysis is valid for impact velocities that are small compared with the smallest acoustic speed associated with wave propagation normal to the specimen face. Since the temperature changes during the experiment are thought to be relatively unimportant, the analysis is carried out in an isothermal setting.

Details concerning the experiments can be found in Escobar [3], Escobar and Clifton [4, 5]. They also develop a visco-plastic model that captures many features of their observations. The material in the present chapter is an abbreviated version of the work described in Abeyaratne and Knowles [2].

In Section 13.2 we set out some preliminary information pertaining to the material and the motion of the specimen. The loading is characterized by a velocity $\mathbf{v}_0 = \varepsilon \mathbf{w}$ where $\mathbf{w}$ is a vector of constant (but not unit) magnitude that characterizes the direction of impact, and the nondimensional parameter ε denotes the ratio between the speed of impact and the smallest relevant acoustic speed. In order to describe some of the mathematical details in the simplest possible setting, in Section 13.3 we formulate and solve the impact problem when the loading does *not* induce a phase transformation. Then in Section 13.4 we analyze the problem with phase transformation. The analysis up to this point is not limited to cubic austenite and monoclinic martensite. In the final section of this chapter we specialize and apply the results to Escobar and Clifton's experiments and finally determine the kinetic relation associated with the austenite $\rightarrow \beta_1'$-martensite transformation in the material they tested. Many details are not described here and the reader is referred to [2] for these. We omit, for example, issues such as how the analysis for a half-space is used to model a thin plate; how the experiment, which involved the impact of one CuAlNi plate on another, can be modeled by the application of a constant velocity on a single plate; and so on.

13.2 Preliminaries

Consider a two-phase material and take the reference configuration to be one composed of unstressed austenite, which is thus characterized by $\mathbf{F} = \mathbf{I}$. Let $\mathbf{F} = \mathbf{F}^0$ denote one particular variant of martensite, also in an unstressed state. By assumption, austenite and martensite both correspond to local minima of the Helmholtz free energy function $\psi(\mathbf{F})$ where, since we are modeling the problem in an isothermal setting, we have suppressed ψ's dependency on temperature. Therefore

$$\psi_{\mathbf{F}}(\mathbf{I}) = \psi_{\mathbf{F}}(\mathbf{F}^0) = \mathbf{0}, \tag{13.1}$$

and the two elasticity tensors

$$\mathbf{C}^+ = \rho\psi_{\mathbf{FF}}(\mathbf{I}), \qquad \mathbf{C}^- = \rho\psi_{\mathbf{FF}}(\mathbf{F}^0) \tag{13.2}$$

are both positive definite. The tensors $\mathbf{I}$ and $\mathbf{F}^0$ lie at the respective bottoms of the austenitic and one martensitic wells. As noted in Section 12.2, if there is to be a phase boundary separating austenite with deformation gradient $\mathbf{I}$ from martensite with deformation gradient $\mathbf{F}^0$, their difference, $\mathbf{F}^0 - \mathbf{I}$, must be a rank-1 tensor. As discussed in Chapter 12, this requirement typically imposes certain restrictions on the lattice parameters, and for most alloys, these conditions do not hold. However in certain alloys, the composition can be selected so as to force these conditions to be satisfied, in which event it is possible to have a piecewise homogeneous two-phase deformation. The alloy tested by Escobar and Clifton is believed to be special in this way and so we assume that $\mathbf{F}^0 - \mathbf{I}$ is rank-1, and therefore that there are vectors $\mathbf{b}$ and $\mathbf{m}$ such that

$$\mathbf{F}^0 - \mathbf{I} = \mathbf{b} \otimes \mathbf{m}; \tag{13.3}$$

$\mathbf{m}$ is normal to the phase boundary. Finally we assume that austenite is the more stable phase at the temperature underlying the isothermal calculation and so

$$\psi(\mathbf{I}) < \psi(\mathbf{F}^0). \tag{13.4}$$

Consider a half-space $x_1 > 0$ composed of austenite, which is initially undeformed and at rest. Its free surface $x_1 = 0$ is subjected to a constant velocity $\mathbf{v}_0 = \varepsilon\mathbf{w}$ for all times $t > 0$. Here ε is a nondimensional parameter and $\mathbf{w}$ has the dimension of velocity. In light of the homogeneity of the material and loading, it is natural to assume that the body undergoes a motion that depends solely on the single spatial coordinate x_1 and time. Thus it is natural to set $x = x_1$ and to consider motions of the form

$$\mathbf{y} = \mathbf{y}(x, t) = \mathbf{x} + \mathbf{u}(x, t) \quad \text{for } x > 0, \ t > 0 \tag{13.5}$$

subject to the initial conditions

$$\mathbf{u}(x, 0) = \mathbf{0}, \qquad \mathbf{u}_t(x, 0) = \mathbf{0} \quad \text{for } x > 0 \tag{13.6}$$

and the boundary condition

$$\mathbf{u}_t(0, t) = \mathbf{v}_0 = \varepsilon\mathbf{w} \quad \text{for } t > 0. \tag{13.7}$$

The vector $\mathbf{m}$ in (13.3) is normal to the phase boundary. If we want the phase boundary to be parallel to the planar surface of the specimen, then we must have $\mathbf{m} = \mathbf{e}_1$ where $\mathbf{e}_1$ is a unit vector that is normal to the plane face of the half-space. By cutting the specimen at the appropriate orientation, one can arrange for the crystal lattice to be oriented relative to the specimen in such a way that this is true. When the lattice is oriented this way we have

$$\mathbf{F}^0 - \mathbf{I} = \mathbf{b} \otimes \mathbf{e}_1. \tag{13.8}$$

It is now possible to have a two-phase equilibrium configuration of the body involving a planar phase boundary that is parallel to the free surface of the half space; the body is then unstressed and is composed of austenite and martensite at deformation gradient tensors $\mathbf{I}$ and $\mathbf{F}^0$ respectively. The vector $\mathbf{b}$ is determined by the underlying crystallography as in Example 5 of Section 12.2; since $\mathbf{b}$ is a measure of the difference between the austenite tensor $\mathbf{I}$ and the martensite tensor $\mathbf{F}^0$ we refer to it as the transformation strain vector.

13.3 Impact without phase transformation

Suppose that the impact parameter ε is small and that the resulting motion does not carry the body far from its reference configuration. Then it is natural to assume that

$$\mathbf{u}(x, t) = \varepsilon \mathbf{u}^+(x, t) + O(\varepsilon^2) \quad \text{for } x > 0, \ t > 0, \tag{13.9}$$

and to linearize the problem to leading order in ε. The reason for the symbol "+" above $\mathbf{u}$ will become clear in the next section when we consider the case where a phase transformation *does* occur, in which case the displacement field has the form (13.9) *ahead* of a moving phase boundary. Throughout the motion under consideration here, the body remains in the austenite phase at all times. To leading order the deformation gradient tensor associated with (13.5) and (13.9) is

$$\mathbf{F} = \mathbf{I} + \varepsilon \, (\mathbf{u}_x^+ \otimes \mathbf{e}_1), \tag{13.10}$$

and the first Piola–Kirchhoff stress tensor $\boldsymbol{\sigma} = W_\mathbf{F}(\mathbf{F})$ approximates to

$$\boldsymbol{\sigma} = \boldsymbol{\sigma}^+ = \varepsilon \mathbf{C}^+ (\mathbf{u}_x^+ \otimes \mathbf{e}_1) \tag{13.11}$$

where $\mathbf{C}^+$ is the positive definite elasticity tensor associated with undistorted austenite defined in (13.2).

To leading order, the Lagrangian equation of motion Div $\boldsymbol{\sigma} = \rho \mathbf{u}_{tt}$ reduces to

$$\mathbf{A}^+ \mathbf{u}_{xx}^+ = \mathbf{u}_{tt}^+ \quad \text{for } x > 0, \ t > 0, \tag{13.12}$$

which is to be solved subject to the initial conditions

$$\mathbf{u}^+ = \mathbf{u}_t^+ = \mathbf{0} \quad \text{for } x > 0, \ t = 0 \tag{13.13}$$

and the boundary condition

$$\mathbf{u}_t^+ = \mathbf{w} \quad \text{for } x = 0, \ t > 0. \tag{13.14}$$

Here the acoustic tensor $\mathbf{A}^+$ associated with undistorted austenite is defined by

$$\mathbf{A}^+ \mathbf{g} = \frac{1}{\rho} [\mathbf{C}^+ (\mathbf{g} \otimes \mathbf{e}_1)] \mathbf{e}_1 \quad \text{for all vectors } \mathbf{g}. \tag{13.15}$$

Note that $\mathbf{A}^+$ is symmetric, positive definite and is independent of both x and t.

In order to solve the linear initial–boundary value problem (13.12)–(13.14), it is convenient to introduce the vector fields

$$\gamma(x, t) = \mathbf{u}_x^+(x, t), \qquad \mathbf{v}(x, t) = \mathbf{u}_t^+(x, t), \tag{13.16}$$

that denote strain and particle velocity respectively. The hyperbolic differential equation (13.12) can involve shock waves. If $\alpha(t)\mathbf{e}_1$ denotes the velocity of such a propagating discontinuity, then the basic kinematic and momentum jump conditions (6.9) and (6.10) can be shown to reduce, in the present setting, to

$$\alpha[\![\gamma]\!] + [\![\mathbf{v}]\!] = \mathbf{0}, \qquad \mathbf{A}^+[\![\gamma]\!] = \alpha^2[\![\gamma]\!]. \tag{13.17}$$

Since the acoustic tensor $\mathbf{A}^+$ is constant, symmetric, and positive definite, it follows from $(13.17)_2$ that the shock propagation speed α is constant, that there are three possible speeds of propagation, and that they are given by the square roots of the eigenvalues of $\mathbf{A}^+$. In the present linearized setting shock wave speeds and acoustic wave speeds coincide. We will assume that the eigenvalues α_k^2, $k = 1, 2, 3$, of $\mathbf{A}^+$ are distinct and denote the associated orthonormal eigenvectors by $\mathbf{a}^{(k)}$:

$$\mathbf{A}^+\mathbf{a}^{(k)} = \alpha_k^2\mathbf{a}^{(k)}, \quad k = 1, 2, 3, \quad \alpha_1 > \alpha_2 > \alpha_3 > 0. \tag{13.18}$$

We now seek a solution of the impact problem in the piecewise constant form corresponding to Figure 13.1:

$$\gamma(x, t), \mathbf{v}(x, t) = \begin{cases} \mathbf{0}, \quad \mathbf{0}, \quad \text{for } x > \alpha_1 t, \\[6pt] \gamma^{(1)}, \ \mathbf{v}^{(1)} \text{ for } \alpha_1 t < x < \alpha_2 t, \\[6pt] \gamma^{(2)}, \ \mathbf{v}^{(2)} \text{ for } \alpha_2 t < x < \alpha_3 t, \\[6pt] \gamma^{(3)}, \ \mathbf{v}^{(3)} \text{ for } 0 < x < \alpha_3 t, \end{cases} \tag{13.19}$$

where $\gamma^{(k)}$, $\mathbf{v}^{(k)}$, $k = 1, 2, 3$, are constant vectors that are to be determined. This solution involves three shock waves at $x = \alpha_k t$, $k = 1, 2, 3$. The initial conditions are automatically satisfied by $(13.19)_1$.

By enforcing the jump conditions (13.17) at each shock wave and making use of (13.18), one readily finds that

$$\left.\begin{aligned} \gamma^{(1)} &= \beta_1\mathbf{a}^{(1)}, \\ \gamma^{(2)} &= \beta_1\mathbf{a}^{(1)} + \beta_2\mathbf{a}^{(2)}, \\ \gamma^{(3)} &= \beta_1\mathbf{a}^{(1)} + \beta_2\mathbf{a}^{(2)} + \beta_3\mathbf{a}^{(3)}, \end{aligned}\right\} \tag{13.20}$$

$$\left.\begin{aligned} \mathbf{v}^{(1)} &= -\alpha_1\beta_1\mathbf{a}^{(1)}, \\ \mathbf{v}^{(2)} &= -\alpha_1\beta_1\mathbf{a}^{(1)} - \alpha_2\beta_2\mathbf{a}^{(2)}, \\ \mathbf{v}^{(3)} &= -\alpha_1\beta_1\mathbf{a}^{(1)} - \alpha_2\beta_2\mathbf{a}^{(2)} - \alpha_3\beta_3\mathbf{a}^{(3)}, \end{aligned}\right\} \tag{13.21}$$

where β_1, β_2, β_3 are three scalars that remain to be determined.

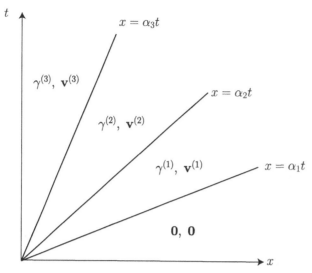

Figure 13.1. The x, t-plane. The three rays $x = \alpha_i t$, $i = 1, 2, 3$, correspond to three shock waves.

Finally we turn to the boundary condition (13.14), which, because of (13.19)$_4$, can be written as

$$\mathbf{v}^{(3)} = \mathbf{w}. \tag{13.22}$$

Substituting (13.21)$_3$ into (13.22) yields a vector equation, which can be solved for the three scalars β_k giving

$$\beta_k = -\frac{1}{\alpha_k}\, \mathbf{w} \cdot \mathbf{a}^{(k)}, \quad k = 1, 2, 3. \tag{13.23}$$

The dissipation inequality (6.33) and (7.13) should be enforced on each shock wave. It is not difficult to show by a direct calculation that the rate of dissipation at a shock wave is $O(\varepsilon^3)$; thus, at shock waves, this inequality imposes no requirements to the order to which we are working.

Thus in summary, when no phase transformation occurs, the response to impact involves three elastic waves propagating into the undisturbed body at speeds α_1, α_2, and α_3. The associated solution is found as follows: given the impact velocity $\mathbf{v}_0 = \varepsilon\mathbf{w}$, the parameters β_k are determined from (13.23), the $\gamma^{(k)}$s and $\mathbf{v}^{(k)}$s are given by (13.20) and (13.21), and then the strain and particle velocity fields in the body are given by (13.19).

13.4 Impact with phase transformation

Now consider the case when the impact loading (13.7) causes a stress-induced transformation from austenite to martensite. Specifically, suppose that a planar phase boundary is nucleated at the impact face at the instant of impact and that it then propagates into the half-space. The region ahead of the phase boundary is associated with austenite, while the region behind it is associated with martensite;

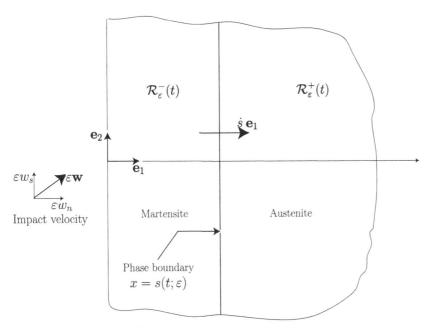

Figure 13.2. Half-space with propagating phase boundary.

see Figure 13.2. Let

$$x = s(t; \varepsilon) \quad \text{with} \quad s(0; \varepsilon) = 0 \tag{13.24}$$

denote the position of the phase boundary at time $t \geq 0$; based on the preceding analysis, one expects three shock waves to run ahead of the phase boundary in the austenitic region. The regions $\mathcal{R}_\varepsilon^+(t)$ and $\mathcal{R}_\varepsilon^-(t)$, ahead of and behind the phase boundary, are characterized by

$$\mathcal{R}_\varepsilon^+(t) = \{\mathbf{x} \mid x > s(t; \varepsilon)\}, \qquad \mathcal{R}_\varepsilon^-(t) = \{\mathbf{x} \mid 0 < x < s(t; \varepsilon)\}. \tag{13.25}$$

Suppose that the impact parameter ε is small. We assume that in the resulting motion the deformation gradient tensor field on the austenitic region does not depart significantly from the identity $\mathbf{I}$, and that on the martensitic region it remains close to the tensor $\mathbf{F}^0$; recall that the respective tensors $\mathbf{I}$ and $\mathbf{F}^0$ characterize austenite and the relevant martensite variant, both unstressed. Consequently we are assuming that during the motion the deformation gradient tensor at any point in the body does not depart significantly from the bottoms of the two associated energy wells even though it may jump between these two wells. More precisely, we assume that the motion is of the form

$$\mathbf{y} = \begin{cases} \mathbf{x} + \varepsilon \mathbf{u}^+(x, t) + \mathbf{c}(t) + O(\varepsilon^2) & \text{on} \quad \mathcal{R}_\varepsilon^+(t), \\[2mm] \mathbf{F}^0 \mathbf{x} + \varepsilon \mathbf{u}^-(x, t) + O(\varepsilon^2) & \text{on} \quad \mathcal{R}_\varepsilon^-(t), \end{cases} \tag{13.26}$$

where $\mathbf{c}(t)$ denotes a rigid translation. Moreover, we suppose that the phase boundary location can also be expanded in a series as

$$s(t; \varepsilon) = s_0(t) + \varepsilon s_1(t) + O(\varepsilon^2). \tag{13.27}$$

The strategy is to linearize the problem on each phase, and to then connect the two regions across the phase boundary using the jump conditions, also appropriately approximated. We omit the details of this calculation and refer the interested reader to [2]. One first finds that $s_0(t) = 0$ and $\mathbf{c}(t) = \mathbf{0}$ for $t \geq 0$; and then to $O(\varepsilon)$, the basic problem to be solved can be described solely in terms of the austenitic displacement field $\mathbf{u}^+(x, t)$ as follows: we are to determine $\mathbf{u}^+(x, t)$ by solving the partial differential equation

$$\mathbf{A}^+ \mathbf{u}_{xx}^+ = \mathbf{u}_{tt}^+ \quad \text{for } x > 0, \ t > 0, \tag{13.28}$$

subject to the initial conditions

$$\mathbf{u}^+ = \mathbf{u}_t^+ = \mathbf{0} \quad \text{for } x > 0, \ t = 0, \tag{13.29}$$

and the boundary condition

$$\mathbf{u}_t^+ = \mathbf{w} + \dot{s}_1 \mathbf{b} \quad \text{for } x = 0, \ t > 0, \tag{13.30}$$

where $\mathbf{b}$ was introduced in (13.3).

Observe that the initial–boundary value problem (13.28)–(13.30) here is formally identical to the problem (13.12)–(13.14) describing the case without phase transformation except that $\mathbf{w} + \dot{s}_1 \mathbf{b}$ in the boundary condition (13.30) was previously $\mathbf{w}$ in (13.14). Thus the strain and particle velocity fields in the impact problem *with* phase transformation can be written down immediately: they are given by (13.19), where the $\gamma^{(k)}$s and $\mathbf{v}^{(k)}$s are given by (13.20) and (13.21), with the parameters β_k now given by

$$\beta_k = -\frac{1}{\alpha_k} [\mathbf{w} + \dot{s}_1 \mathbf{b}] \cdot \mathbf{a}^{(k)}, \quad k = 1, 2, 3. \tag{13.31}$$

This provides the solution to the impact problem in terms of the unknown phase boundary speed, which, to leading order, is $\varepsilon \dot{s}_1$.

If the kinetic relation of the phase transformation was known, then we could complete the solution by using the kinetic law to determine $\dot{s}_1$. However in this chapter we take the point of view that the kinetic relation is *not* known but that certain experimental data are available instead. In the next section we show how experimental measurements can be used to calculate $\dot{s}_1$, thus completing the solution. The kinetic law relating the propagation speed $V_n = \varepsilon \dot{s}_1$ to the driving force f on the phase boundary is also determined using the data.

Table 13.1. Measured data from Table 4.1 of Escobar [3]. Reproduced with permission from JoAnne Escobar.

Experiment	Normal impact velocity εw_n (m s^{-1})	Tangential impact velocity $-\varepsilon w_s$ (m s^{-1})	Transverse particle velocity $-\varepsilon \mathbf{v}^{(3)} \cdot \mathbf{e}_2$ (m s^{-1})
1	21.15	17.75	14.50
2	29.60	16.75	13.60
3	43.65	26.20	12.65
4	46.60	39.10	15.85

13.5 Application to austenite-β_1' martensite transformation in CuAlNi

13.5.1 Experimental data. As mentioned previously, the specimen was cut relative to the cubic lattice of the austenite such that $\mathbf{F}^0 - \mathbf{I} = \mathbf{b} \otimes \mathbf{e}_1$ where $\mathbf{e}_1$ is the unit normal to the free surface that points into the specimen; this ensures the existence of an austenite–martensite interface that is parallel to the face of the specimen. In order to facilitate the initiation of the phase transformation, the impact velocity $\mathbf{v}_0 = \varepsilon \mathbf{w}$ was chosen to lie in the plane defined by the specimen normal $\mathbf{e}_1$ and the transformation strain vector $\mathbf{b}$. Let $\{\mathbf{e}_1, \mathbf{e}_2, \mathbf{e}_3\}$ be an orthonormal basis with $\mathbf{e}_2$ chosen so that both $\mathbf{b}$ and $\mathbf{v}_0$ lie in the plane defined by $\mathbf{e}_1$ and $\mathbf{e}_2$. Then if εw_n is the speed of normal impact and εw_s is the speed of tangential impact, we have

$$\mathbf{v}_0 = \varepsilon \mathbf{w} = \varepsilon w_n \mathbf{e}_1 + \varepsilon w_s \mathbf{e}_2. \tag{13.32}$$

Escobar [3] has reported the data in Table 13.1 from four impact experiments, giving the normal and tangential components of the impact velocity and the transverse component of the resulting particle velocity $\varepsilon \mathbf{v}^{(3)}$ behind the third shock wave $x = \alpha_3 t$.

We shall omit most details concerning the measurements and the calculations of the specific numerical values of various quantities; this information can be found in [2]. The material parameters measured or estimated by various investigators include the lattice parameters η_1, η_2, η_3, the mass density ρ, the elastic moduli of austenite C_{ijkl}^+, the transformation temperature θ_T, and the latent heat at the transformation temperature λ_T. Among the calculations omitted here are one where we use (12.33) to calculate the transformation strain vector $\mathbf{b}$ and another where we use (13.15) to calculate the acoustic tensor $\mathbf{A}^+$ for austenite and its eigenvalues $\alpha_1, \alpha_2, \alpha_3$ and eigenvectors $\mathbf{a}^{(1)}, \mathbf{a}^{(2)}, \mathbf{a}^{(3)}$. The three acoustic speeds come out to be

$$\alpha_1 = 5767 \, \text{m s}^{-1}, \qquad \alpha_2 = 3553 \, \text{m s}^{-1}, \qquad \alpha_3 = 1187 \, \text{m s}^{-1}, \tag{13.33}$$

and therefore the values of the parameter ε, which is the ratio between the impact speed and the smallest acoustic speed, are

$$\varepsilon = 0.023, \quad 0.029, \quad 0.043, \quad 0.051, \tag{13.34}$$

in the four tests.

Table 13.2. Results: phase boundary speed.

Experiment	Propagation speed $\dot{s}\,(\text{m}\,\text{s}^{-1})$
1	19.21
2	18.63
3	80.42
4	137.92

13.5.2 Phase boundary speed. We now calculate the *phase boundary speed* $\dot{s} = \varepsilon \dot{s}_1$. Note first from $(13.21)_3$ and (13.31) that $\mathbf{v}^{(3)} = \mathbf{w} + \dot{s}_1 \mathbf{b}$. Thus $\dot{s}_1$ can be calculated by taking the dot product of this equation with any vector that is not orthogonal to $\mathbf{b}$. Since the transverse particle velocity $\varepsilon \mathbf{v}^{(3)} \cdot \mathbf{e}_2$ was measured, it is natural to take the dot product of this with $\mathbf{e}_2$ leading to

$$\dot{s} = \varepsilon \dot{s}_1 = \varepsilon \frac{(\mathbf{v}^{(3)} - \mathbf{w}) \cdot \mathbf{e}_2}{\mathbf{b} \cdot \mathbf{e}_2} = \frac{\varepsilon \mathbf{v}^{(3)} \cdot \mathbf{e}_2 - \varepsilon w_s}{\mathbf{b} \cdot \mathbf{e}_2} \qquad (13.35)$$

from which $\dot{s}$ can be calculated. The resulting values of $\dot{s}$ in each experiment are shown in Table 13.2.

13.5.3 Driving force. We turn next to the driving force $f = \rho \psi^+ - \rho \psi^- - \frac{1}{2}(\sigma^+ + \sigma^-) \cdot (\mathbf{F}^+ - \mathbf{F}^-)$, which can be linearized to yield

$$f(t) = \rho \psi(\mathbf{I}) - \rho \psi(\mathbf{F}^0) + \varepsilon \rho \mathbf{b} \cdot [\mathbf{A}^+ \mathbf{u}_x^+(s_0(t), t)] + O(\varepsilon^2),$$

$$= \rho \psi(\mathbf{I}) - \rho \psi(\mathbf{F}^0) + \varepsilon \rho \sum_{k=1}^{3} \alpha_k^2 \beta_k \, \mathbf{a}^{(k)} \cdot \mathbf{b} + O(\varepsilon^2), \qquad (13.36)$$

where we have also used $(13.16)_1$, $(13.19)_4$, $(13.20)_3$, and (13.18) in obtaining the second expression.

In order to estimate the first two terms let

$$f_o(\theta) = \rho \psi(\mathbf{F}^0, \theta) - \rho \psi(\mathbf{I}, \theta) \qquad (13.37)$$

where θ is the test temperature. Since $f_o(\theta)$ represents the height between the two energy wells at the temperature θ, it must vanish at the temperature θ_T, be positive for $\theta > \theta_T$, and be negative for $\theta < \theta_T$. If we approximate $f_o(\theta)$ by the first two terms of its Taylor series and use the fact that $f_o(\theta_T) = 0$, we get

$$f_o(\theta) \approx (\rho \psi_\theta(\mathbf{F}^0, \theta_T) - \rho \psi_\theta(\mathbf{I}, \theta_T))(\theta - \theta_T). \qquad (13.38)$$

Next, recall that the specific entropy per unit mass, η, is related to ψ, the Helmholtz free energy per unit reference volume, by $\eta(\mathbf{F}, \theta) = -\psi_\theta(\mathbf{F}, \theta)$; recall also that the latent heat per unit mass of the austenite $\rightarrow$ martensite transformation, at the temperature θ_T, is given by $\lambda_T = -\theta_T[\eta(\mathbf{F}^0, \theta_T) - \eta(\mathbf{I}, \theta_T)]$. On combining the

Table 13.3. Results: driving force.

Experiment	Driving force f (MPa)
1	25.36
2	25.51
3	26.54
4	31.81

above relationships, we obtain the following rough estimate for f_o:

$$f_o \approx \rho \lambda_T \left(\frac{\theta}{\theta_T} - 1 \right). \tag{13.39}$$

Thus the value of the driving force in each test can be calculated from (13.36), (13.37), and (13.39). The results are shown in Table 13.3.

13.5.4 Kinetic law. In Figure 13.3, we plot the four pairs $(f, \dot{s})$ given in Tables 13.2 and 13.3 that correspond to the experiments. These points should lie on or near the curve corresponding to the kinetic relation $\dot{s} = \Phi(f)$ characterizing the austenite to β_1'-martensite transformation. The solid curve in Figure 13.3 is the graph of the function

$$\Phi(f) = \begin{cases} 0 & \text{for} \quad f \leq f_*, \\ M\sqrt{f^2 - f_*^2} & \text{for} \quad f \geq f_*, \end{cases} \tag{13.40}$$

with $M = 7.07\,\text{m}\,\text{MPa}^{-1}\,\text{s}^{-1}$ and $f_* = 25.26\,\text{MPa}$. Kinetic response functions of this form have been shown in [1] to describe the kinetics of twinning of γ_1'-martensite in the same CuAlNi alloy used by Escobar and Clifton. For further discussion, see Chapter 14 of this monograph or Abeyaratne, Chu and James [1].

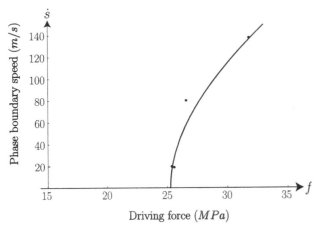

Figure 13.3. Kinetic law. The four points in the figure correspond to the data in Tables 13.2 and 13.3. The curve is described by equation (13.40) with $M = 7.07\,\text{m}\,\text{MPa}^{-1}\,\text{s}^{-1}$ and $f_* = 25.26$ MPa.

REFERENCES

[1] R. Abeyaratne, C. Chu, and R.D. James, Kinetics of materials with wiggly energies: theory and application to the evolution of twinning microstructures in a Cu–Al–Ni shape memory alloy. *Philosophical Magazine* A, **73** (1996), pp. 457–97.

[2] R. Abeyaratne and J.K. Knowles, On the kinetics of an austenite→martensite phase transformation induced by impact in a Cu–Al–Ni shape memory alloy. *Acta Materialia*, **45** (1997), pp. 1671–83.

[3] J.C. Escobar, *Plate Impact Induced Phase Transformations in Cu–Al–Ni Single Crystals*. Ph.D. dissertation, Brown University, Providence, RI, 1995.

[4] J.C. Escobar and R.J. Clifton, On pressure-shear plate impact for studying the kinetics of stress-induced phase transformations. *Journal of Materials Science and Engineering*, **A170** (1993), pp. 125–42.

[5] J.C. Escobar and R.J. Clifton, Pressure-shear impact-induced phase transitions in Cu–14.4 Al–4.19 Ni single crystals. *SPIE*, **2427** (1995), pp. 186–97.

14 Quasistatics: Kinetics of Martensitic Twinning

14.1 Introduction

In the preceding chapter we determined the kinetics of a certain phase transformation using experiments that involved fast loading in which inertia was important. In the present chapter we determine the kinetics of a different transformation using data from quasistatic experiments. The transformation studied here is a twinning deformation, not a phase transformation, a twin boundary being an interface that separates two variants of martensite; see Example 1 in Section 12.2. The change in lattice orientation across a twin boundary makes it analogous, in *certain* ways, to a phase boundary, and in particular, the motion of a twin boundary is governed by a kinetic relation.

As we have seen, the simplest form of kinetic relation governing the isothermal motion of an interface relates the driving force on it to its normal velocity of propagation: $V_n = \Phi(f)$. Since the kinetic response function Φ here is a function of a single scalar independent variable, one set of experiments, say uniaxial tension tests, completely determines Φ; and the function Φ thus determined characterizes all motions of this interface such as, say, in biaxial conditions. If the deformation field is inhomogeneous, and the phase or twin boundary is curved, one would use this same kinetic relation locally, at each point along the interface, relating the driving force at that point to the normal velocity of propagation of that point. Consequently once the kinetic relation $V_n = \Phi(f)$ is known, one ought to be able to, at least in principle, calculate the motion of an interface under very general conditions.

One of the goals of the present chapter is to point out that the behavior of interfaces is *not always that simple*. For example, sometimes an interface can get "pinned" at some point along its surface resulting in that point being held back while the rest of the interface propagates according to the kinetic law. In order to understand the motion of an interface in such circumstances, one must be able to characterize the process of pinning/unpinning, in addition to knowing the kinetic relation for normal propagation.

Often, one is interested only in the macroscopic behavior of the material at a scale that homogenizes the motion of individual interfaces and phenomena such as

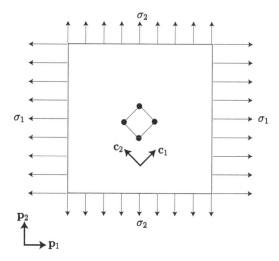

Figure 14.1. Biaxially stressed specimen with orientation of cubic lattice shown.

pinning/unpinning. This chapter illustrates one way to get at this *effective kinetic behavior* directly.

14.2 The material and loading device

The effective kinetic relation that we develop here uses the experiments conducted by Chu and James [4] in which they subjected a thin square plate, roughly 1 in × 1 in × 0.5 mm, to biaxial dead-loading; for example, see Figure 1 of [2]. A schematic sketch of the system is shown in Figure 14.1. If $\mathbf{p}_1$ and $\mathbf{p}_2$ denote unit vectors aligned with the two long edges of the plate, the uniform dead-loading applied on the edges can be described by the Piola–Kirchhoff traction $\mathbf{s} = \sigma\mathbf{n}$ where

$$\sigma = \sigma_1\mathbf{p}_1 \otimes \mathbf{p}_1 + \sigma_2\mathbf{p}_2 \otimes \mathbf{p}_2; \tag{14.1}$$

$\sigma_1 \geq 0$ and $\sigma_2 \geq 0$ are the two stress components applied by the testing machine.

The plate was made of a Cu–14.44Al–4.19Ni (wt%) shape memory alloy, an alloy that we have encountered in previous chapters of this monograph. In particular, the schematic phase diagram in Figure 5.1 indicates that this alloy has three phases and that a different phase is preferred under different conditions of stress and temperature. For example the high stresses generated in the impact experiment described in Chapter 13 caused the specimen there to transform into the monoclinic phase β_1'-martensite. The experiments relevant to the present discussion were conducted under conditions where the loaded specimen remained in the orthorhombic phase γ_1'-martensite at all times.

The γ_1'-martensite lattice can be obtained from the cubic austenite lattice as follows: let $\{\mathbf{c}_1, \mathbf{c}_2, \mathbf{c}_3\}$ be a set of orthonormal vectors in the cubic directions. Stretch the cube by a stretch ratio η_3 in, say, the $\mathbf{c}_3$ direction, and stretch it by η_1 and η_2 along the diagonals of the face normal to $\mathbf{c}_3$. The resulting mapping from the cubic to the orthorhombic phase is $\mathbf{x} \to \mathbf{U}_1\mathbf{x}$ where $\underline{U}_1$, the matrix of components of $\mathbf{U}_1$, is given by

$$\underline{U}_1 = \begin{pmatrix} \frac{\eta_1+\eta_2}{2} & \frac{\eta_1-\eta_2}{2} & 0 \\ \frac{\eta_1-\eta_2}{2} & \frac{\eta_1+\eta_2}{2} & 0 \\ 0 & 0 & \eta_3 \end{pmatrix} \qquad (\text{variant}-1). \qquad (14.2)$$

Here and henceforth, all components of vectors and tensors are taken with respect to the cubic basis $\{\mathbf{c}_1, \mathbf{c}_2, \mathbf{c}_3\}$. The stretch tensor $\mathbf{U}_1$ describes one of the martensitic variants. There are six such variants associated with the three different cube edges and the two associated cube-face diagonals. The second variant associated with stretching the cube by η_3 in the $\mathbf{c}_3$ direction is described by

$$\underline{U}_2 = \begin{pmatrix} \frac{\eta_1+\eta_2}{2} & \frac{\eta_2-\eta_1}{2} & 0 \\ \frac{\eta_2-\eta_1}{2} & \frac{\eta_1+\eta_2}{2} & 0 \\ 0 & 0 & \eta_3 \end{pmatrix} \qquad (\text{variant}-2). \qquad (14.3)$$

In the experiments described in [2, 4] one edge, say $\mathbf{c}_3$, of the cubic lattice was normal to the face of the specimen, and the other two were rotated by $\pi/4$ with respect to the specimen edges as shown in Figure 14.1. For this orientation, the two variants (14.2) and (14.3) were energetically preferred over the four other variants and so the others were not encountered during any of the experiments. Therefore we shall not display the matrices $\underline{U}_i$, $i = 3, \ldots, 6$, associated with the remaining four variants.

The Helmholtz free energy function $\psi(\mathbf{F})$ has an energy well at every tensor $\mathbf{F}$ that has the representation $\mathbf{F} = \mathbf{QU}_i$ for any $i = 1, 2, \ldots, 6$ and every proper orthogonal tensor $\mathbf{Q}$. Without loss of generality we can take the Helmholtz free energy to vanish at the bottom of each of the martensite energy wells: $\psi(\mathbf{QU}_i) = 0$ where $i = 1, 2, \ldots, 6$ and $\mathbf{Q}$ is proper orthogonal.

14.3 Observations

As noted already, for the particular specimen orientation used in the experiments, variants-1 and -2 were energetically preferred. When the applied stress $\sigma_1 \gg \sigma_2$, the entire specimen was composed of variant-1; when $\sigma_2 \gg \sigma_1$ it was composed of variant-2. During a typical test, the biaxial stress histories $\sigma_1(t), \sigma_2(t)$ were prescribed and the specimen transformed between these two variants.

During the process of transformation, the specimen typically involves a mixture of the two variants in the form of a "laminate" – a family of alternating bands of variant-1 and variant-2 as shown schematically in Figure 14.2; the dark bands in the figure correspond to variant-1 and the white bands between them correspond to variant-2. The interfaces between adjacent bands, that is, the boundaries separating one variant from the next, are "twin boundaries."

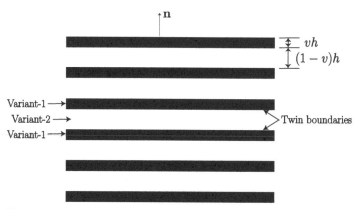

Figure 14.2. A single laminate: dark bands of variant-1 and white bands of variant-2. The volume fraction of variant-1 is v.

As the applied loading varies, the twin boundaries move normal to themselves, and so the bands of one variant get thicker while those of the other variant get narrower. Thus the volume fractions of the variants change. During a typical test, the volume fraction history $v(t)$ of variant-1 arising in the center of a laminate was measured. The response to various loading histories $\sigma_1(t)$, $\sigma_2(t)$ was studied.

Typically, the specimen involves a patchwork of many laminates as shown schematically in Figure 14.3, where each laminate consists of a family of alternating bands of variant-1 and variant-2 as in Figure 14.2. One laminate is separated from an adjacent laminate by a "transition layer" T_h. The bands of one laminate terminate in the transition layer when they meet the bands of an adjacent laminate.

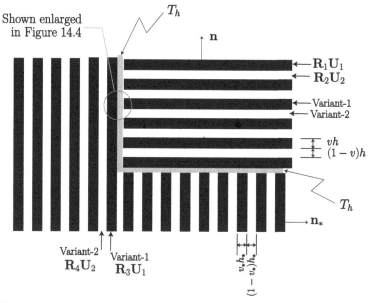

Figure 14.3. Two laminates. The transition layer between laminates is T_h.

We shall not describe the numerous experimental observations of Chu and James but simply remark on one of them. Suppose that variant-1 is, theoretically, the energetically preferred variant at some stress level $(\sigma_1, \sigma_2) = (\sigma_1^*, \sigma_2^*)$. During the experiments, subjecting the specimen to (σ_1^*, σ_2^*) sometimes yielded a configuration involving a mixture of both variants *rather than* the energetically favored one involving only the single preferred variant. Even more significantly, suppose that the specimen is subjected to a gradually varying cyclic loading with the stresses $(\sigma_1(t), \sigma_2(t))$ oscillating around the value (σ_1^*, σ_2^*); specifically, suppose that the amplitude of the oscillation decreases monotonically and that it eventually goes to zero. Even for this loading, the specimen continued to yield a two-variant microstructure rather than one composed entirely of the preferred variant. This is typical of many shape memory alloys and is indicative of the system getting "stuck" in some metastable state that prevents it from achieving its minimum energy configuration. This phenomenon manifested itself in various ways in the different experiments described in [2, 4]. These observations suggest that a naive model of the kinetics might not adequately describe the response of the system.

The goal of the present chapter, which summarizes the work reported in [1, 2], is to understand the appropriate kinetics for this problem and to develop a model that predicts the evolution of the volume fraction $v(t)$ corresponding to a given loading history $(\sigma_1(t), \sigma_2(t))$.

14.4 The model

Since we are concerned here with the effective kinetic response of the material at a scale that sees individual twin bands in a homogenized manner, we construct an internal variable model. Internal variable models were briefly alluded to previously in Sections 3.2 and 3.5. Let E denote the total energy per unit reference volume of the specimen associated with a deformation $\mathbf{y}(\mathbf{x})$. This comprises the elastic energy of the specimen and the energy of the loading device:

$$E = \frac{1}{\text{vol}(\mathcal{R})} \left[\int_{\mathcal{R}} \rho\psi(\nabla\mathbf{y}(\mathbf{x}))d\mathbf{x} - \int_{\partial\mathcal{R}} \boldsymbol{\sigma}\mathbf{n} \cdot \mathbf{y}d\mathbf{x} \right] \qquad (14.4)$$

where $\mathcal{R}$ is the region occupied by the specimen in the reference configuration. Given the stress (σ_1, σ_2) applied by the loading device, suppose that the specimen attains a certain configuration involving variants-1 and -2, with the volume fraction of variant-1 being v. Then the total energy of the system associated with this particular microstructure can, in principle, be calculated using (14.4) and the resulting energy will have the form

$$E = E(v, \sigma_1, \sigma_2). \qquad (14.5)$$

The volume fraction v therefore plays the role of an internal variable and the driving force f conjugate to v is

$$f = -\frac{\partial E}{\partial v}. \qquad (14.6)$$

In the formalism of internal variable theory, the evolution of the system is governed by a kinetic relation relating, for example, the rate of change of the internal variable to the driving force: $\dot{v} = \Phi(f)$. For small values of driving force we can replace it by its linearized approximation

$$\dot{v} = \mathsf{M}f = -\mathsf{M}\frac{\partial E}{\partial v}, \tag{14.7}$$

as long as the mobility $\mathsf{M} = \Phi'(0) \neq 0$. Equation (14.7) is often referred to as a *gradient flow* of the energy E. We are thus led to the following initial value problem for determining $v(t)$ corresponding to given $\sigma_1(t)$, $\sigma_2(t)$:

$$\dot{v}(t) = -\mathsf{M} E'\left(v(t), \sigma_1(t), \sigma_2(t)\right), \quad v(0) = v_0, \tag{14.8}$$

where we have set $E' = \partial E/\partial v$.

Thus the task at hand is to calculate the energy of the system $E(v, \sigma_1, \sigma_2)$ and then study the kinetics as described by (14.8).

14.5 The energy of the system

The total energy of the system comprises two parts, the elastic energy of the specimen and the energy of the loading device, each of which we now calculate separately.

14.5.1 Elastic energy of the specimen. The first term in (14.4) represents the energy E_{elastic} associated with the specimen. In calculating it, it is convenient to consider the transition layers T_h and the rest of the specimen $\mathcal{R} - T_h$ separately. First consider the transition layers T_h. It has been shown by Ball and James [3] that the volume of the transition layers becomes vanishingly small as the bands become narrower: $\mathrm{vol}(T_h) \to 0$ as $h \to 0$. Thus, keeping in mind that the deformation gradient tensor field is uniformly bounded, we can neglect the effect of the transition layers in calculating the elastic energy of the specimen provided that the martensitic bands are very fine.

Next consider the region $\mathcal{R} - T_h$. Here the twin boundaries are essentially straight and parallel as sketched in Figure 14.3 and the deformation is more or less piecewise homogeneous. The strain at a generic particle is due, in part, to elastic deformation and, in part, to transformation. If a particle $\mathbf{x}$ is in variant-i and its deformation gradient tensor is $\mathbf{F}(\mathbf{x})$, the elastic strain of this particle is the "distance" in deformation gradient space of $\mathbf{F}(\mathbf{x})$ from the energy well associated with variant-i. On the other hand the transformation strain is the distance between the two energy wells $|\mathbf{U}_1 - \mathbf{U}_2|$. From a knowledge of the lattice parameters η_1, η_2, η_3 the latter can be estimated to be $|\mathbf{U}_1 - \mathbf{U}_2| = [\,\mathrm{tr}\,(\mathbf{U}_1 - \mathbf{U}_2)^2]^{1/2} \approx 0.055$. On the other hand since the material has a modulus of about 50 GPa and the specimen encounters a typical stress level of about 15 MPa in the experiments, the elastic strain is of the order of $15/50 \times 10^{-3} \approx 0.0003$. Therefore the elastic strains are 200 times smaller than the transformation strain.

When the specimen is unstressed, the deformation gradient tensor $\nabla\mathbf{y}(\mathbf{x})$ at a particle $\mathbf{x}$ lies at the bottom of an energy well. As the specimen is stressed, it moves off the bottom. The preceding estimation indicates that the distance it moves away from this energy well is much smaller than the distance it "jumps" when it eventually transforms to the other energy well. Therefore we shall neglect the former (the elastic strains) in comparison with the latter (the transformation strain) and assume that *the deformation gradient tensor always lies at the bottom of an energy well*.

However, as observed previously, the Helmholtz free energy can be taken to vanish at the bottom of each martensite energy well. It now follows that $\psi(\nabla\mathbf{y}(\mathbf{x})) = 0$ at each $\mathbf{x} \in \mathcal{R} - T_h$ and therefore that the elastic energy associated with $\mathcal{R} - T_h$ vanishes.

Thus in summary, we estimate the elastic energy in both T_h and $\mathcal{R} - T_h$ to vanish and so take the total elastic energy of the specimen to be zero: $E_{\text{elastic}} = 0$.

14.5.2 Loading device energy. The second term in (14.4) represents the energy E_{loading} associated with the dead loading device. In view of the continuity of the deformation and the uniform traction boundary condition, this term can be written as

$$E_{\text{loading}} = -\frac{1}{\text{vol}(\mathcal{R})}\boldsymbol{\sigma} \cdot \left[\int_{\mathcal{R}} \nabla\mathbf{y}(\mathbf{x})d\mathbf{x} \right] = -\boldsymbol{\sigma} \cdot \bar{\mathbf{F}} \qquad (14.9)$$

where $\bar{\mathbf{F}}$ is the average deformation gradient in the specimen and the biaxial stress $\boldsymbol{\sigma}$ is prescribed by (14.1).

In order to calculate $\bar{\mathbf{F}}$, we consider the transition layers T_h and the rest of the specimen $\mathcal{R} - T_h$ separately. First consider the transition layers. As noted above, $\text{vol}(T_h) \to 0$ as $h \to 0$, and the deformation gradient tensor field is uniformly bounded. Thus, provided that the martensitic bands are very fine, one can neglect the effect of the transition layers in calculating the average deformation gradient in the specimen.

Therefore in calculating the average deformation gradient tensor $\bar{\mathbf{F}}$ we may ignore T_h and treat the specimen as a collection of laminates, where each laminate involves a family of parallel bands of variants-1 and -2 as sketched in Figure 14.3. The deformation within each band is homogeneous. In any one laminate, the deformation gradient tensor has the form $v\mathbf{R_1}\mathbf{U_1} + (1 - v)\mathbf{R_2}\mathbf{U_2}$ where $\mathbf{R_1}$ and $\mathbf{R_2}$ are rotations and v is the volume fraction of variant-1 in that laminate (Figure 14.3). Continuity of the deformation across a twin boundary within that laminate requires that $\mathbf{R_1}\mathbf{U_1} - \mathbf{R_2}\mathbf{U_2}$ be a rank-1 tensor and this, effectively, allows one of the rotations, say $\mathbf{R_2}$, to be determined; see the proposition by Ball and James and the surrounding discussion in Section 12.1. In an adjacent laminate, since the same variants are involved, the stretch tensors are again $\mathbf{U}_1$ and $\mathbf{U}_2$; however, the rotations $\mathbf{R}_3$, $\mathbf{R}_4$ and the volume fraction v_* may be different. Continuity of the deformation within this laminate requires that $\mathbf{R_3}\mathbf{U_1} - \mathbf{R_4}\mathbf{U_2}$ be a rank-1 tensor and this again effectively determines one of these $\mathbf{R}$'s, say $\mathbf{R}_4$. Finally, continuity of the deformation between laminates requires that the average deformation gradient tensors in the two laminates, $v\mathbf{R_1U_1} + (1 - v)\mathbf{R_2U_2}$ and $v_*\mathbf{R_3U_1} + (1 - v_*)\mathbf{R_4U_2}$, differ by a rank-1 tensor. This allows one to determine one of the two unknown rotations, say $\mathbf{R_3}$, and one of the volume fractions, say v_*. No matter how many laminates are involved, one can proceed systematically in this way and determine all of the rotations and volume fractions in terms of a single volume fraction v in a particular laminate and a single rotation $\mathbf{R}$.

Therefore the loading device energy takes the form $E_{\text{loading}} = -\boldsymbol{\sigma} \cdot \bar{\mathbf{F}} = -\boldsymbol{\sigma} \cdot$ $\mathbf{R}D(v)$ where $\boldsymbol{\sigma}$ is prescribed and $\mathbf{D}$ is known in terms of v. The two unknowns in this formula for E_{loading} are the overall rotation of the specimen $\mathbf{R}$ and the volume fraction v. During a loading process, the former is determined by the rigid body dynamics of the specimen and the latter by the transformation kinetics. Assuming that the macroscopic dynamics takes place much faster than the kinetics as was true in the experiments, we choose $\mathbf{R}$ such that it minimizes E_{loading} at fixed v and $\boldsymbol{\sigma}$. This leads to the following explicit expression for E_{loading} in terms of the applied stress (σ_1, σ_2), the volume fraction v, and the lattice parameters η_1, η_2, η_3:

$$E_{\text{loading}}(v, \sigma_1, \sigma_2) = -\left\{ v^2 \frac{(\eta_1^2 - \eta_2^2)^2}{(\eta_1^2 + \eta_2^2)} (\sigma_1^2 + \sigma_2^2) \right.$$
$$\left. + 2v \frac{(\eta_1^2 - \eta_2^2)}{(\eta_1^2 + \eta_2^2)} (\sigma_1^2 \eta_2^2 - \sigma_2^2 \eta_1^2) + (\sigma_1 \eta_2 + \sigma_2 \eta_1)^2 \right\}^{\frac{1}{2}}.$$

$$(14.10)$$

The details of the calculations leading to this result, together with a careful description of the underlying assumptions, can be found in [2].

14.5.3 Summary. In summary, according to the preceding calculations the total energy of the system is $E = E_{\text{elastic}} + E_{\text{loading}} = E_{\text{loading}}$ where the elastic energy vanishes according to Section 14.5.1 and the loading device energy is given by (14.10)[1].

One can readily verify from (14.10) that E_{loading} is concave in the volume fraction whence

$$\frac{\partial^2 E}{\partial v^2}(v, \sigma_1, \sigma_2) < 0 \qquad (14.11)$$

on the appropriate range for the variables. Therefore the minimum of $E(\cdot, \sigma_1, \sigma_2)$ occurs at either $v = 0$ or $v = 1$ indicating that a single variant state has less energy than a two-variant state associated with a laminate. Moreover, one can verify that this energy is minimized by $v = 1$ when $\sigma_1 > \sigma_2$ indicating that the specimen prefers to be composed of variant-1 when $\sigma_1 > \sigma_2$. Conversely the energy is minimized

---

[1] The relative importance of these two energetic contributions was examined in [2], where it was found that the loading device energy (14.10) is about 400 times as large as the nominal elastic energy in the specimen, $1/2(\text{stress}^2/\text{modulus})$, which justifies the approximation $E_{\text{elastic}} \ll E_{\text{loading}}$.

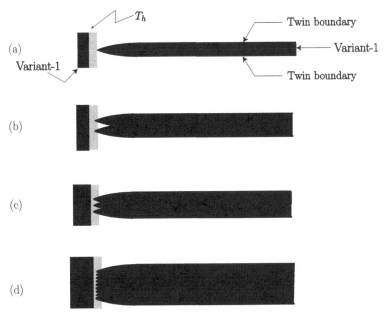

Figure 14.4. An enlarged view of the region enclosed by the circle in Figure 14.3. The figure shows a time sequence of sketches of a single band of one variant in one laminate meeting a band of the same variant in a neighboring laminate. The motion of the band is constrained by the pinning of one or more tips. As the width of the band increases, the tips split and their number increases.

by $v = 0$ when $\sigma_2 > \sigma_1$ so that the specimen prefers to be composed of variant-2 in this case. This is consistent with the observations described at the beginning of Section 14.3.

Turning to the kinetics, the inadequacy of our model becomes readily apparent. Consider for example a creep test in which the system evolves with the applied stress (σ_1, σ_2) held fixed. Because of the concavity of E, the solution of (14.8) necessarily tends (as $t \to \infty$) to $v = 0$ if the initial volume fraction $v(0)$ is less than v_* and to $v = 1$ if $v(0) > v_*$ where v_* is given by $\partial E(v_*, \sigma_1, \sigma_2)/\partial v = 0$. This contradicts the experimental observations described previously where (a) two-variant laminates are often seen in equilibrium (i.e. $v(\infty) \neq 0$ or 1) and (b) many two-variant equilibrium states are found to exist at the same (σ_1, σ_2). This indicates that our model is too naive and forces us to reexamine our calculation of the energy more closely.

14.6 The effect of the transition layers: Further observations

Figure 14.4 shows an enlarged view of the region enclosed by the circle in Figure 14.3. This is the region in which one laminate meets another. Consider a single band of variant-1 as shown schematically in Figure 14.4(a). This thin and long band is contained between two twin boundaries, which are parallel to each other over most of its length. However as shown in the figure, at the very ends of the band, the two boundaries curve sharply toward each other and pinch down into a tip.

These tips are pinned: they constrain the motion of the twin boundary because, as the parallel portions of the twin interfaces move normal to themselves and the band gets wider, the tips remain more or less stationary. As the band gets wider, at some critical band width, the tip splits into two or more tips, and the band width increases discontinuously. This process repeats itself with the width continuing to increase with accompanying bursts of additional tip-splitting as described by the sequence of schematic diagrams progressing from (a) to (d) in Figure 14.4.

The phenomenon of tip-splitting is particularly striking during a creep test. Suppose that the specimen settles into an equilibrium configuration. Consider a single band, which has a certain width and some number of tips. It is found that even a very slight disturbance affects this microstructure. For example, lightly tapping the specimen with a pencil causes additional splitting, and the number of tips and the width of the band changes, *even though* the applied stresses σ_1, σ_2 have not changed. This can be repeated by further tapping.

A configuration with a certain number of tips corresponds to a particular metastable state of the band. This same band, at the same load, has other metastable states available to it, each corresponding to a certain number of tips. The aforementioned phenomenon of getting stuck in metastable states is intimately connected to the availability of these different configurations. Thus in order to properly characterize the evolution of the volume fraction in the specimen, one must account for both the kinetics of the normal motion of the twin interfaces *and* the process of tip-splitting.

In calculating the total energy of the system, we must therefore account for the energy associated with the transition layers T_h in which the tip-splitting occurs. Based on the preceding discussion, this energy should have a large number of local minima, where each minimum corresponds to a particular configuration of a band involving a certain number of tips. The energy barriers between the minima should be small, since the observations show that it is easy to pass from one metastable state, corresponding to a certain number of tips, to another, associated with a different number of tips.

14.7 The effect of the transition layers: Further modeling

We now calculate the energy associated with the transition layers. There are two contributions to this energy: the elastic energy stored within the region T_h and the energy associated with tip-splitting. In contrast to the regions away from the transition layers where the deformation is essentially piecewise homogeneous, the deformation in the transition layers is highly inhomogeneous due to the fading away of the twin bands as they enter this region. The elastic energy in T_h can be estimated using ideas of Kohn and Müller [5, 6], which leads to

$$E_{\text{elastic}} = c_1 v^2 + c_2 (1 - v)^2. \tag{14.12}$$

We omit the derivation of (14.12) and refer the reader to [2] for details. The constants c_1 and c_2 are expected to depend on the elastic moduli, the surface energy density, and the dimensions of the laminates ([5, 6]).

The relative importance of the various energetic terms was examined in [2], where it was found that the loading device energy (14.10) is about 50 times as large as the elastic energy as given by (14.12). However even though the elastic energy in the transition layer is 50 times smaller than the loading device energy, it plays a crucial role in light of the fact that it affects the convexity of the total energy. In particular, for suitable values of the stress, the function $E(\cdot, \sigma_1, \sigma_2) = E_{\text{elastic}}(\cdot, \sigma_1, \sigma_2) + E_{\text{loading}}(\cdot, \sigma_1, \sigma_2)$ is minimized by values of volume fraction v, which are *between* 0 and 1 corresponding to microstructures involving a mixture of variants; this is in contrast to the loading device energy E_{loading}, which is always minimized by $v = 0$ or 1.

As observed in Section 14.6, the energy associated with the phenomenon of tip-splitting must have a large number of local minima, each corresponding to a particular configuration of a band with a certain number of tips. The energy barriers between the minima should be small since the observations show that it is easy to increase the number of tips by additional splitting, that is, it is easy to pass from one metastable state to another. Moreover, the distance between two adjacent minima must be small since the change in volume fraction accompanying the appearance of a new tip is small. These conditions suggest that we model the energy associated with the tips by an expression such as

$$E_{\text{tips}}(v, \sigma_1, \sigma_2) = a\varepsilon \cos\left(\frac{v}{\varepsilon}\right), \tag{14.13}$$

where a and ε are constants with $\varepsilon \ll 1$. The height of the energy barriers is $2a\varepsilon$ and the distance between adjacent minima is $2\pi\varepsilon$.

14.8 Kinetics

The total energy of the system is now given by

$$E_\varepsilon(v, \sigma_1, \sigma_2) = E_{\text{elastic}} + E_{\text{loading}} + E_{\text{tips}} \tag{14.14}$$

where the energies E_{elastic}, E_{loading}, and E_{tips} are given by (14.12), (14.10), and (14.13) respectively. The kinetic relation is now a gradient flow based on (14.14) and this leads to the initial-value problem

$$\dot{v}_\varepsilon(t) = -M\, E_\varepsilon'(v_\varepsilon(t), \sigma_1(t), \sigma_2(t)), \qquad v_\varepsilon(0) = v_0 \tag{14.15}$$

for determining the volume fraction $v_\varepsilon(t)$ corresponding to given $\sigma_1(t)$, $\sigma_2(t)$; here $E_\varepsilon' = \partial E_\varepsilon/\partial v$. The existence of multiple local minima of $E_\varepsilon(\cdot, \sigma_1, \sigma_2)$ suggests that this evolution law does have the capability for describing the phenomenon of getting stuck in metastable states.

Since we are interested in the macroscopic response of the specimen and not in the detailed behavior of each individual twin band, we need to homogenize the evolution law (14.15) by passing to the limit $\varepsilon \to 0$. The mathematical analysis of the homogenization of (14.15) will not be presented here; this can be found in [2]. The analysis in [2] proves that the solution $v_\varepsilon(t)$ of (14.15) converges uniformly to

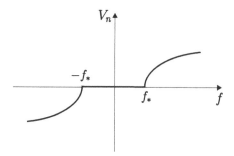

Figure 14.5. Macroscopic kinetic law.

$v(t)$ where $v(t)$ obeys the *macroscopic kinetic law*

$$\dot{v} = \begin{cases} \mathsf{M}\sqrt{f^2 - f_*^2}, & f > f_*, \\ 0, & -f_* < f < f_*, \\ -\mathsf{M}\sqrt{f^2 - f_*^2}, & f < -f_*, \end{cases} \qquad v(0) = v_o. \qquad (14.16)$$

Here $f = -\partial\hat{E}/\partial v$ is the driving force calculated from the energy $\hat{E} = E_{\text{elastic}} + E_{\text{loading}}$, which does *not* include tip-splitting effects; and $f_* = a$ where $a\varepsilon$ is the amplitude of the tip-splitting energy (14.13). Equation (14.16) shows that for small values of driving force, $|f| < f_*$, the volume fraction is "stuck" and does not evolve and that it begins to change only when the driving force is sufficiently large. We note in passing that this phenomenon of getting stuck (often called "lattice trapping") is common to other models of kinetics as well; see, for example, the Frenkel–Kontorowa model in Section 8.4.5.

The model constructed here involves four unknown parameters c_1, c_2, M, and f_*, the lattice parameters η_1, η_2, η_3 having been measured independently. One set of numerical values of these four parameters was used in [2] to show that the quantitative predictions of this kinetic law compare reasonably well with the numerous responses measured by Chu and James; see [2].

REFERENCES

[1] R. Abeyaratne, C. Chu, and R.D. James, Kinetics and hysteresis in martensitic single crystals, in *Proceedings of Symposium on the Mechanics of Phase Transformations and Shape Memory Alloys*. American Society of Mechanical Engineers Applied Mechanics Division, 1994, Vol. 189, pp. 85–98.

[2] R. Abeyaratne, C. Chu, and R.D. James, Kinetics of materials with wiggly energies: theory and application to the evolution of twinning microstructures in a Cu–Al–Ni shape memory alloy. *Philosophical Magazine* A, **73** (1996), pp. 457–97.

[3] J.M. Ball and R.D. James, Fine phase mixtures as minimizers of energy. *Archive for Rational Mechanics and Analysis*, **100** (1987), pp. 13–52.

[4] C. Chu, *Hysteresis and Microstructures: A Study of Biaxial Loading on Compound Twins of Copper–Aluminum–Nickel Single Crystals*. Ph.D. dissertation, University of Minnesota, Minneapolis, MN, 1993.

[5] R.V. Kohn and S. Müller, Branching of twins near an austenite–twinned-martensite interface. *Philosophical Magazine* A, **66** (1992), pp. 697–715.

[6] R.V. Kohn and S. Müller, Surface energy and microstructure in coherent phase transitions. *Communications on Pure and Applied Mathematics*, **47** (1994), pp. 405–35.

Author Index

Subject Index